Cost Analysis for Engineers and Scientists

Manufacturing and Production Engineering Series

Series Editors: Hamid R. Parsaei, Texas A&M University at Qatar, & Waldemar Karwowski, University of Central Florida

This series will provide an outlet for the state-of-the-art topics in manufacturing and production engineering disciplines. This new series will also provide a scientific and practical basis for researchers, practitioners, and students involved in areas within manufacturing and production engineering. Issues envisioned to be addressed in this new series would include, but not limited to, the following: Additive Manufacturing, 3D Visualization, Mass Customization, Material Processes, Cybersecurity, Data Science, Automation and Robotics, Underwater Autonomous Vehicles, Unmanned Autonomous Vehicles, Robotics and Automation, Six Sigma and Total Quality Management, Manufacturing Cost Estimation and Cost Management, Industrial Safety, and Programmable Logic Controllers, to name just a few.

Decision Making in Risk Management
Quantifying Intangible Risk Factors in Projects
Christopher Cox

Reconfigurable Manufacturing Enterprises for Industry 4.0
Ibrahim H. Garbie and Hamid Parsaei

Recent Advances in Time Series Forecasting
Dinesh C. S. Bisht and Mangey Ram

Cost Analysis for Engineers and Scientists
Fariborz (Fred) Tayyari

For more information on this series, please visit: https://www.routledge.com/Manufacturing-and-Production-Engineering/book-series/CRCMNPRDENG

Cost Analysis for Engineers and Scientists

Fariborz (Fred) Tayyari

CRC Press
Taylor & Francis Group
Boca Raton London New York

CRC Press is an imprint of the
Taylor & Francis Group, an **informa** business

First edition published 2022
by CRC Press
6000 Broken Sound Parkway NW, Suite 300, Boca Raton, FL 33487-2742

and by CRC Press
2 Park Square, Milton Park, Abingdon, Oxon, OX14 4RN

Library of Congress Cataloging-in-Publication Data

Names: Tayyari, F. (Fariborz), author.
Title: Cost analysis for engineers and scientists / Fariborz Tayyari.
Description: First edition. | Boca Raton, FL : CRC Press, 2022. | Series: Manufacturing and production engineering | Includes bibliographical references and index.
Identifiers: LCCN 2021027979 (print) | LCCN 2021027980 (ebook) | ISBN 9781138362284 (hbk) | ISBN 9781032147116 (pbk) | ISBN 9780429432163 (ebk)
Subjects: LCSH: Engineering economy.
Classification: LCC TA177.4 .T38 2022 (print) | LCC TA177.4 (ebook) | DDC 658.15--dc23
LC record available at https://lccn.loc.gov/2021027979
LC ebook record available at https://lccn.loc.gov/2021027980

ISBN: 978-1-138-36228-4 (hbk)
ISBN: 978-1-032-14711-6 (pbk)
ISBN: 978-0-429-43216-3 (ebk)

DOI: 10.1201/9780429432163

Typeset in Times
by Deanta Global Publishing Services, Chennai, India

Access the Support Material: https://www.routledge.com/9781138362284

Contents

Preface

Cost analysis is a powerful managerial tool that requires a systematic analysis that results in a better understanding and budgeting costs of products and services. This understanding, in turn, leads to improved program management in applying resources and mitigating project risks.

This book is intended to empower engineers and scientists with skills for obtaining needed financial data, and to apply cost analysis techniques for assessing product and service costs, and costs associated with engineering projects. The readers will learn how to keep records of financial transactions, prepare and analyze financial statements, and interpret them to make sound decisions. They also develop skills for manufacturing cost determination and estimation, overhead allocation, cost estimation, and profitability analysis. The subject materials covered in this book help cost engineers and scientists apply managerial accounting techniques to problems in the areas of cost estimation, cost control, product pricing, product line or business segment discontinuation, profitability analysis, and project management.

This book is intended as a valuable resource for training students and learners who will be part of a team that is responsible for the execution of industrial projects. The contents of this book are aimed at all individuals who are involved in technical projects, cost estimation, cost control and planning, and contract review in their organizations, including but not limited to:

- Organization managers.
- Division/department directors.
- Production managers.
- Project managers.
- Cost estimators.
- Cost analysts and engineers.

The book, first, introduces the fundamentals of accounting information systems and manufacturing costs. Then, it presents product costing, manufacturing cost allocation to individual products and joint products, and manufacturing cost estimation. Finally, it presents the concepts and applications of cost-volume-profit and break-even analysis for a single product and multiple products, as well as single period and multiple periods.

Fariborz (Fred) Tayyari

Preface

Author biography

Fariborz (Fred) Tayyari is a professor of Industrial & Manufacturing Engineering & Technology at Bradley University. He holds a bachelor's degree in Cost Accounting, Master of Business Administration (MBA), Master of Science in Industrial Engineering (MSIE), and a PhD in Industrial Engineering. He has been teaching Engineering Economy, Engineering Costs Analysis, and Operations Research for more than 35 years. He also served a major technical and construction company as chief accountant and as a budget officer during his military service. He received a certification as a Professional Engineer (PE) from the state of Texas, a Certified Professional Ergonomist (CPE), and a Certified Safety Professional (CSP). He has published book chapters on topics such as Total Cost Analysis of Modern Automated Systems and Joint Cost Allocation.

Author Biography

1 Accounting and Cost Information Systems

LEARNING OBJECTIVES

- Learn the basics of the accounting process and information.
- Maintain accounting bookkeeping and records.
- Prepare financial statements for a newly established business.
- Recognize classes of assets, liabilities, and net worth.
- Understand the profitability of a business in financial reports, which are sources of financial decision-making for both internal and external individuals.

1.1. INTRODUCTION

A meaningful engineering economic and cost analysis needs familiarity with financial information. Accounting records provide such information. In order for the engineering economics analysts to be able to efficiently communicate with accountants and obtain the necessary data, they need to understand accounting terminologies and the accounting and cost information systems.

Regardless of the type and size of a business, accounting information is crucial for decision-making, financial planning, and assessing economic performance.

This chapter introduces some basic accounting terminologies, accounting principles, and accounting concepts. Learning the terms and concepts, the reader will be able to understand the financial reports and statements and other features of accounting and cost information systems. Knowledge of such information is crucial for engineers and scientists who seek economic justification for their engineering proposals or career in management.

1.2. TERMS AND DEFINITIONS

Accounting: The American Institute of Certified Public Accountants [1] defined accounting as follows:

> *"Accounting is the art of recording, classifying and summarizing in a significant manner and in terms of money, transactions and events which are, in part at least, of a financial character and interpreting the results thereof."*

DOI: 10.1201/9780429432163-1

This definition helps the reader better understand accounting through the following basic interpretations:

- **Accounting is an art**: Accounting is an art because it requires skills and creative judgment. A trained accountant performs accounting functions.
- **Accounting involves interconnected "phases"**: ***Recording*** is the process of writing down or keeping records of business transactions. ***Classifying*** involves grouping similar items that have been recorded. The classified data are *summarized* and reported in financial statements.
- **Accounting is only concerned with *transactions and events having financial character***: For example, purchase of materials or goods, sale of goods, the payment of salaries, payment of taxed or insurance premium, and acquisition of a facility are recorded because they have *monetary* values. However, hiring an engineer and signing a service contract is just qualitative information without financial characteristics. Hence, they are not recorded in the accounting books.
- ***Interpreting the results***: The data in the financial statements and reports have meanings that are useful to the interested parties. Interpreting results is part of *accounting functions* (the final phase of accounting).

Accounting records are sources of cost information systems for measuring, processing, and communicating financial data that are valuable in making economic decisions. In general, accounting practices include bookkeeping, preparation of financial statements, and reports on all business activities (e.g., change in assets, sales revenues and expenses, inventories, receivables, payables, and payments of bills). However, the accounting functions widely vary among industries (e.g., manufacturing, construction, service, and governments) and even among firms within each industry.

Bookkeeping: Bookkeeping is the means of recording monetary transactions and keeping records.

Cost Accounting: Cost accounting refers to the determination and control of costs. Cost accountants deal with the costs of manufactured products as well as the expenses of selling and distributing the products.

Account: An account is a basic data storage unit used in accounting. Accountants use one account for each asset, liability, and components of owner's equity (capital), revenue, and expense.

Chart of Accounts: The chart of accounts is a listing of all accounts used in the accounting books (i.e., general ledger) of an organization. An accounting system uses a chart to aggregate information into an entity's financial statements. Accountants sort the chart by account numbers to ease the task of locating specific accounts. See Appendix 1.A.

Journal: A journal (or general journal) is a chronological record of all monetary transactions. It is the first place for recording the transactions.

Ledger: A ledger is a book or all files of a company's accounts.

Depreciation: Depreciation cost or expense is a periodic allocation of the cost of a tangible, long-lived asset over its estimated useful life.

Cost of Goods Sold (COGS): Cost of goods sold (or costs of sales) is the "**cost of goods available for sale**" during that period minus the cost of inventory (merchandise) at the end of the period.

Cost of Goods Manufactured: Cost of goods manufactured represents the total manufacturing costs assigned to units of a product manufactured during a period.

Operating Expenses: Operating expenses are nonmanufacturing expenses incurred due to normal operations of a business (e.g., selling and administrative expenses).

Nonoperating Expenses: Nonoperating expenses are expenses that are not related to normal business operations, such as interest expenses incurred due to borrowing money.

Double-Entry System. The field of accounting – both the older manual systems and today's computerized systems – is based on an old accounting procedure known as double entry, which has been practiced for several centuries. Double entry is a simple yet powerful concept, based on which, the accountants must record each monetary transaction into at least two of the accounts in the accounting books, whereby the amounts entered as debits must be equal to the amounts entered as credits.

T-Account: T-account is a simple form of an account that has a physical resemblance to the letter "T," used to analyze transactions. As shown in Figure 1.1, a T-account has the following three parts:

- A title that describes the account (e.g., Cash, Accounts Receivable, Notes Receivable, Inventories, Building, Equipment, Accounts Payable, Notes Payable, Taxes Payable, Mortgage Payable, Common Stocks, Preferred Stocks, and Retained Earnings).
- A left side, called the "debit" side.
- A right side, called the "credit" side.

Debit and **Credit**: The terms debit (abbreviated Dr., from the Latin ***debere***) and credit (abbreviated Cr., from the Latin ***credere***) are two words used by accountants to, respectively, show the left and right sides of an account. The following rules for debit and credit are applied to increase or decrease an account:

- A debited entry to an asset account increases that asset, while a credited entry decreases the asset.

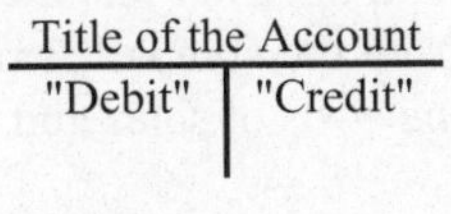

FIGURE 1.1 An illustration of a T-account.

ASSETS		=	LIABILITIES		+	OWNER'S EQUITY	
Debit to increase	Credit to decrease		Debit to decrease	Credit to increase		Debit to decrease	Credit to increase

FIGURE 1.2 The effects of debits and credits on account balances.

- A credited entry to a liability or owner's equity account increases that liability or owner's equity account, while a debited entry decreases the liability or owner's equity account.

These rules can be summarized as shown in Figure 1.2.

Notice that revenues increase the owner's equity (to which the rule of owner's equity is applied), and expenses decrease the owner's equity (to which the opposite rule of owner's equity is applied, that is, treated like assets).

Revenues: Revenues are the total amount received for the goods sold or services provided to customers during a particular fiscal (financial or accounting) period. The company receives either cash or a promise to receive cash in the future for its sales or services to the customers. Revenues increase assets (Cash, Accounts Receivable, and/or Notes Receivable), and finally result in an increase in the owner's equity.

Expenses: Expenses are the costs of activities performed to carry out the company's operations. Salaries to office employees and sales personnel, commissions, telephone services, utilities, office rent, advertising, and office supplies are some examples of expenses. Expenses decrease assets and/or increase liabilities. They are the opposite of revenue in that they finally result in a decrease in owner's equity.

Accrued Expenses: Accrued expenses are expenses that a firm incurs in one accounting period but will pay for it in another period. Common examples of accrued expenses are salaries and wages payable, taxes payable, interest payable, and warranty services on products performed, all of which have been incurred, but for which a firm has not received invoices or has received close to the end of the fiscal period, and the firm has not paid them in that period.

Accrued Revenues: Accrued revenues are revenues that a firm has earned in one fiscal period but will receive payment for them in another period. The balance of the "accounts receivable" account is the accrued revenues from sales of products or services, and the balance of "dividends and interest receivable" account is the accrued dividends or interests income in a fiscal period but the firm will receive payments for them in another period.

Net Income and **Net Loss**: The term net income refers to the net increase in owner's equity resulting from the company's profitable operations. It is the end result of subtracting the costs of sales and expenses from revenues:

$$\textbf{Net Income}\left(\textbf{Loss}\right) = \textbf{Revenues} - \textbf{Costs of Goods Sold and Expenses}$$

The term "net loss" refers to the net decrease in owner's equity due to the company's profit-seeking operations that result in a negative net income. Net loss occurs whenever the company's revenues are not enough to recover all of its cost of goods sold and expenses.

Prepaid Expenses: Prepaid expenses are advance payments for services or goods that a firm will receive in the near future. Prepaid expenses appear on accounting books and reports as assets. When the firm receives the goods or services, the prepaid expenses are expired, and the firm will convert them into expenses. Prepaid insurance and prepaid rent are the two common examples of prepaid expenses.

Unearned Revenues: Unearned revenues are amounts a firm has collected for goods or services to deliver to its customers in the future (not yet delivered). Just as opposite to prepaid expenses, unearned revenues are advance receipts of money from customers before they are actually earned (or realized). Unearned revenues are obligations of the company to provide goods or services to the customers. Therefore, an unearned revenue (or fee) is a "current" liability. As the company fulfills any portion of these obligations, it earns that portion of the advance receipts. Then, it converts the earned portion from the unearned revenues (a current liability) into earned revenue.

1.3. FINANCIAL STATEMENTS

The financial statements reflect the financial positions and periodical business activities of a firm. The most important financial statements are:

1) Income Statement.
2) Owners' Equity Statement.
3) Balance Sheet.
4) Cash Flow Statement.

These reports are prepared in the given order and presented to the public as a set of financial statements. This means they are not only published together but also designed and intended to be used together. As each statement only gives information about specific aspects of a company's financial activities and position, it is important that the reports are used together.

1.3.1. The Income Statement

An income statement is a financial statement that shows the amount of income earned by a business (say, ABC Company) over an accounting period (e.g., for the year ending December 31, 20XX). Figure 1.3 illustrates an income statement showing the profitability of a hypothetical company during the year 20XX. The following are the examples of income statement accounts:

ABC Company
Income Statement
For the year ended December 31, 20XX

Revenues (Total Sales)		$32,000
Cost of Goods Sold		10,000
Gross Margin		$22,000
Less: Operating Expenses:		
Salaries & Wages	$1,500	
Heat, Light, & Power	1,000	
Communications Expense	200	
Insurance	500	
Depreciation	1,800	
Total Operating Expenses		5,000
Operating Income		$17,000
Less: Non-operating Expenses:		
Interest Expense		1,500
Income Before Income Taxes		$15,500
Income Taxes		7,500
Net Income After Tax		$ 8,000

FIGURE 1.3 An illustration of the income statement.

- **Revenue accounts** – Examples: Sales of Merchandise, Service Revenues, Investment Revenues.
- **Expense accounts** – Examples: Cost of Goods (Merchandise) Sold, Sales Commissions, Advertising Expenses, Wages Expense, Rent Expense, Depreciation Expense.

1.3.2. Owners' Equity Statement

The statement of stockholder's equity, often called the statement of changes in equity, is prepared for a fiscal period (e.g., quarter). This statement displays how equity changes from the beginning of a fiscal period to the end.

The statement of stockholder's equity shows the equity accounts that affect the ending equity balance including common stock, net income, paid-in capital, and dividends. This statement has four sequential sections:

1) Beginning balance.
2) Additions.
3) Subtractions.
4) Ending balance.

The beginning equity balance is followed by two indented sections: additions and subtractions. Additions include new investments and net income if the company is

ABC Company **Statement of Owners' Equity** For the year ended December 31, 20XX	
Beginning Balance, January 1, 20XX	$24,500
Add: Net Income	8,000
Less: Dividends	(2,000)
Ending Balance, December 31, 20XX	$30,500

FIGURE 1.4 An illustration of a statement of owners' equity.

profitable. If the company is not profitable, net loss for the year is included in the subtractions along with any dividends to the owners. The last line on this statement always lists the ending equity balance. Figure 1.4 illustrates the four parts of a statement of owners' equity.

1.3.3. Balance Sheet

The balance sheet shows the financial position of a business on a particular date, usually on the last day of its fiscal period. The balance sheet presents a view of the business as a collection of assets belonging to the company that is equal to the resources or claims against those assets. The following are examples of balance sheet accounts:

- **Asset accounts**
  - **Current Assets** – Examples: Cash, Marketable Securities, Accounts Receivable, Supplies, Prepaid Insurance, and Inventories.
  - **Fixed Assets** – Examples: Land, Buildings, and Equipment.
  - **Intangible Assets** – Examples: Patents, Copyrights, and Goodwill.
- **Liability accounts**
  - **Current Liabilities** – Examples: Accounts Payable, Wages Payable, Taxes Payable, and Short-Term Notes Payable.
  - **Long-Term Liabilities** – Examples: Notes Payable (Long-Term), Bonds (matured in longer than one year), and Mortgage Loans.
- **Stockholders' Equity Accounts** – Examples: Preferred Stock, Common Stock, Paid-In Capital, and Retained Earnings.

The resources consist of liabilities (debts of external parties) and the owner's equity (the owner's interest in the company). An example of the balance sheet for a hypothetical company on December 31, 20XX, is given in Figure 1.5.

Accounting Equation: This is also known as the basic accounting equation and balance sheet equation. As the balance sheet shows, the sum of assets (left side) equals the sum of liabilities and owner's equity (right side), as shown in Figure 1.5. The following is the accounting equation:

$$\textbf{Assets} = \textbf{Liabilities} + \textbf{Owners' Equity}\left(\textbf{or Net Worth}\right)$$

XYZ Company
Balance Sheet
As of December 31, 20XX

ASSETS		LIABILITIES	
Cash	$1,000	Accounts Payable	$3,500
Accounts Receivable	1,200		
Land	600	**OWNERS' EQUITY**	
Building	9,000	Mr. XYZ, Capital	19,300
Equipment	11,000		
Total Assets	**$22,800**	**Total** Liabilities & Owners' Equity	$22,800

FIGURE 1.5 A simple balance sheet.

Equation	Equation Components				
					Owners' (Stockholders') Equity
Basic:	Assets	=	Liabilities	+	Owners' Equity
Expanded:	Assets	=	Liabilities	+	Common Stock + Paid-in Capital – Treasury Stock + Revenues – Expenses – Dividends

FIGURE 1.6 The components of the basic and expanded accounting equations.

Expanded Accounting Equation: This equation splits the equity part of the basic accounting equation into four elements: owner's capital, owner's withdrawals, revenues, and expenses. Both the assets and liabilities sections of the basic equation remain the same in the expanded equation. Figure 1.6 compares the components/elements of the two forms of the equation (basic and expanded) for a corporation.

1.3.4. Cash Flow Statement

The cash flow statement presents a company's cash or cash equivalent inflows and outflows from operations, financing activities, and investing activities. It indicates how well the company generates cash to pay its debt obligations and funds its operating expenses. This statement is a document that shows the actual cash that came in and went out of a firm during a given period from operating activities, investing activities, and financing activities. Figure 1.7 exemplifies the cash flow statement.

The operating activities on the cash flow statement include any sources and uses of cash from business activities. Examples of operating activities include:

- Receipts from sales of goods and services.
- Payments made to suppliers of goods and services used in production.

KBL Company
Cash Flow Statement
For the Period Ended December 31, 20XX

Cash Flow from Operating Activities		
Net Income (Earnings)	$1,200,000	
Depreciation	12,000	
Decrease in Accounts Receivable	20,000	
Decrease in Accounts Payable	(55,000)	
Accrued Expenses	4,000	
Increase in Unearned Income	3,000	
Increase in Taxes Payable	21,000	
Increase in Prepaid Rent	(20,000)	
Increase in Inventory	(25,000)	
***Net Cash from Operating* Activities**		**$1,160,000**
Cash Flow from Investing Activities		
Issuance and Sales of Common Stock	100,000	
Purchase of Equipment	(500,000)	
Renovations and Improvements	(220,000)	
***Net Cash from Investing* Activities**		**(620,000)**
Cash Flow from Financing Activities		
Dividends Payments	(180,000)	
Decrease in Notes Payable	(200,000)	
***Net Cash from Financing* Activities**		**(380,000)**
Change in Cash in period ended Dec. 31, 20XX		**$160,000**
Beginning Cash Balance		**125,000**
Ending Cash Balance, December 31, 20XX		**$285,000**

FIGURE 1.7 An example of a cash flow statement.

- Salary and wage payments to employees.
- Rent payments.
- Insurance premium payments.
- Income tax payments.
- Any other type of operating expenses.

Examples of investing activities include:

- Receipts from sales of marketable securities.
- Interests and revenues (e.g., dividends) received from investment.

Examples of financing activities include:
- Interest payments (on loans and bonds).

1.4. BOOKKEEPING

Bookkeeping is the process of recording transactions and keeping records.

1.4.1. Double-Entry System of Accounting

As previously stated, in the double-entry system, each transaction must be recorded twice (debit and credit) in such a way that the total debits equal total credits. The mechanism of double entry in bookkeeping is such that every transaction affects at least two accounts with the total of debits equals to the total of credits.

Example 1.1

The list below contains the transactions for the H.T. Ram Property Management for the month of January, 20XX.

1. On January 2, H.T. Ram invested $7,500 to establish his property management agency.
2. On January 3, rented an office, paying two months' rent in advance, $700. He also hired an office assistant and agreed to pay $400 every two weeks.
3. On January 4, purchased office equipment from Sam Equipment for $1,750, paying $750 in cash and agreeing to pay the rest next month.
4. On January 5, purchased on credit office supplies for $500 from York Supply Company, and paid $120 for a one-year insurance policy.
5. On January 10, paid York Supply Company $250 of the amount owed.
6. On January 12, performed a service for a customer and collected a fee of $850.
7. On January 16, paid the office assistant two weeks' salary, $400.
8. On January 18, accepted $700 as an advance fee from Mr. Turner (one of his customers).
9. On January 20, performed a service for a customer. The earned fees of $950 will be collected next month.
10. On January 25, Mr. Ram withdrew $960 from the business for personal living expenses.
11. On January 29, paid the office assistant two more weeks' salary, $400.
12. On January 30, paid the utility bill, $85.
13. On January 31, received a telephone bill, $48.

Figure 1.8 illustrates the principle of duality (double entry) and how accountants post the transactions in the affected T-accounts (ledger accounts).

1.4.2. General Journal

Bookkeepers journalize all transactions, chronologically in the "general journal." **Journalizing transactions** is the process of keeping a record of all your business transactions in chronological order, which generally includes the date, the accounts

Cash	
(1) 7,500	700 (2)
(6) 850	750 (3)
(8) 700	120 (4)
	250 (5)
	400 (7)
	960 (10)
	400 (11)
	85 (12)

Accounts Payable	
(5) 250	1,000 (3)
	500 (4)
	48 (13)

H.T. Ram, Capital	
	7,500 (1)

Accounts Receivable	
(9) 950	

Unearned Fees	
	700 (8)

H.T. Ram, Withdrawals	
(10) 960	

Office Supplies	
(4) 500	

Fees Earned	
	850 (6)
	950 (9)

Prepaid Rent	
(2) 700	

Utility Expense	
(12) 85	

Wages (Expense)	
(7) 400	
(11) 400	

Prepaid Insurance	
(4) 120	

Telephone Expense	
(13) 48	

Office Equipment	
(3) 1,750	

FIGURE 1.8 Posting transactions of Example 1.1 in the corresponding T-accounts.

debited and credited, and a brief description of the transactions. Each transaction is recorded through a separate journal entry.

A company may use more than one kind of journal. The simplest and most flexible type is the general journal. The general journal contains the following columns for the recording:

- Date.
- Name of the accounts debited and credited.
- Dollar amounts debited or credited to each account.
- A brief explanation of the transaction (whenever necessary).
- Coding numbers to identify the accounts (known as posting reference).

Example 1.2

XYZ Company purchased $350 office supplies and $1,500 office equipment on January 4 on credit, and prepaid $360 for a one-year insurance policy on January 6. These two transactions are recorded in the general journal as shown in Figure 1.9.

Note that at the time of recording the transactions, nothing is recorded in the "post. ref." (posting reference) column. If the company uses an identifying number for each account in the ledger, the account numbers are placed in this column to provide a cross-reference from the general journal to the ledger and indicate that posting to the ledger has been made.

XYZ Company
General Journal

Page 002

Date		Description	Post. Ref.	Debit	Credit
20XX					
January	5	Office Supplies		350	
		Office Equipment		1,500	
		Accounts Payable			1,850
		Purchase of Office Supplies and equipment on credit			
	6	Prepaid Insurance		360	
		Cash			360
		Paid for one-year insurance premium			

FIGURE 1.9 A partial page of the general journal for Example 1.2.

1.4.3. General Ledger

Accountants, based on recording history, allocate one or more pages of the ledger for each account. The ledger accounts provide the following information for posting transactions (Figure 1.10).

- Name (title) of the account at the top of the page (e.g., Figure 1.10 shows Accounts Receivable).
- Account number, at the top of the page (no. 112 for Accounts Receivable).
- Dates of the transactions.
- Description (or **Item**), usually not used unless special notes are desirable.
- Posting reference to show the page number of the general journal, where the transaction was first recorded.
- Two columns for debiting or crediting the account.
- Two columns for the account balance (debit or credit).

XYZ Company
General Ledger
Accounts Receivable

Account No. 112

Date		Item	Post. Ref.	Debit	Credit	Balance Debit	Balance Credit
20XX						$	$
Jan.	3			500		500	
	5				200	300	
	12			750		1,050	
	25			450		1,500	
Feb.	8				600	900	
	15			100		1,000	

FIGURE 1.10 An illustration of a page (a cut out) of a "general ledger."

1.4.4. Posting Journal Entries into Ledger Accounts

After bookkeepers have journalized the transactions (recorded transactions in the general journal), they must post (transfer) the amounts to the corresponding ledger accounts. The process of transferring the transactions from the journal to the ledger is called posting. They usually perform posting at the end of each day or after several journal entries, depending on the number of transactions. The posting process includes the following steps:

1) Locate in the ledger the account named in the general journal to debit or credit.
2) Enter the date of the transaction in the date column and the journal page number, from which the entry comes, in the post. ref. column of the ledger.
3) Enter the amount of the debit (or credit) as appeared in the journal in the debit (or credit) column of the ledger.
4) Enter the account number, to which the amount was posted, in the posting reference column of the general journal.

The posting reference columns of the general journal and the general ledger are for referencing between the journal entries and the ledger accounts. It indicates the completion of posting of the item. Figure 1.11 shows the relationship between the journal entries and the ledger accounts.

General Journal Page **14**

DATE		DESCRIPTION	Post. Ref.	Debit	Credit
20XX					
July	15	Office Supplies	**122**	565	
		Accounts Payable	**211**		565
		Purchase of Office Supplies on credit			

General Ledger
Office Supplies Account No. **122**

Date		Item	Post. Ref.	Debit	Credit	Balance	
						Debit	Credit
20XX		Balance from previous page				3,885	
July	15		**14**	565		4,450	

General Ledger
Accounts Payable Account No. **211**

Date		Item	Post. Ref.	Debit	Credit	Balance	
						Debit	Credit
20XX		Balance from last page					43,310
July	15		**14**		565		43,875

FIGURE 1.11 The relationship between the journal entries and the ledger accounts.

1.4.5. Terms of Sales

When merchandise is sold on credit, both seller and buyer should have a clear understanding of the terms of sales. The terms are usually printed on the sales invoice. The payment may be expected within a short period of time such as 15 days or 30 days. In such cases, the sales invoice may carry the designation "n/15" or "n/30," meaning that the net sales amount is due 15 days or 30 days, respectively, after the invoice date. Sometimes, the amount is due 10 days after the end of the month (eom). In this case, the invoice is designated "n/10 **eom**."

However, it is common to offer discounts to encourage early payment. In such a case, the designated terms may be stated as "**5/15, n/60**" or "**2/10, n/30**." The terms "5/15, n/60" mean that the customer may take a 5% discount if the customer pays the invoice within 15 days after the invoice date. Otherwise, the customer can delay payment until 60 days after the invoice date and pay the full invoice amount without the discount.

Since the seller is not sure that whether the customer will take the advantage of the discount offer, the sales discounts are recorded on the date the payment is received from the customer within the discount period. For example, assume that MacDowell Fabric Company sells merchandise to Joe Smith, a customer, in the amount of $3,000 on terms of "3/10, n/45." The journal entry by MacDowell Fabric Company at the time of the sale would be as follows:

Nov. 18	Accounts Receivable	3,000	
	Sales		3,000
	Merchandise sold on credit to Joe Smith, terms 3/10, n/45		

If the customer (Joe Smith) takes advantage of the sales discount and pays in ten days or sooner, say on November 26, the journal entry would be as follows:

Nov. 26	Cash	2,910	
	Sales Discount	90	
	Accounts Receivable		3,000
	Receipt of payment from Joe Smith for Invoice of Nov. 18, less 3% discount		

Similar to sales discounts, accountants debit the Sales Returns and Allowances account for the sales items returned by the buyer. At the end of a fiscal period, to determine the net sales, accountants deduct the totals of both the Sales Discount account and Sales Returns and Allowances account from gross sales (balance of Sales account). These two accounts are called sales contra-accounts.

1.5. RECORDING PURCHASES AND SALES OF MERCHANDISE

Firms may adopt either the ***periodic system*** or the ***perpetual system*** to account for their inventories.

1.5.1. Recording Purchases and Sales Using Periodic Inventory System

Under the periodic inventory system, accountants debit the Purchases account for merchandise purchased, and credit the "Purchases Returns and Allowances" and the "Purchases Discount" accounts for returning any part of purchases and getting discounts on the purchases, respectively. The firms update the Inventory account's balance only at the end of each fiscal period. Similarly, they credit the Sales (Revenue) account for merchandise sold, and debit the "Sale Returns and Allowances" and the "Sales Discount" accounts for any part of items sold that are returned by customers and discounts given to them on the sales, respectively. The firms update the Inventory account's balance only at the end of each fiscal period; they, also, debit the "Cost of Goods Sold" and credit inventories for the costs of the items sold as part of inventory updating.

1.5.2. Recording Purchases and Sales Using Perpetual Inventory System

Under the perpetual inventory system, firms record purchases, purchase returns and allowances, purchase discounts, sales, and sales returns at costs in the inventory account. This system maintains the inventory account balance accurate (up-to-date), assuming there are no spoilage or other losses.

Assume that a firm buys and sells gadgets; its several transactions during the month of May of a given year (say 2020) under both systems are as given in Figure 1.12. We have removed the reference columns from the general journal for simplicity purposes and use the acronyms A/R for accounts receivable and A/P for accounts payable.

Traditionally, firms selling small numbers of expensive items have used perpetual inventory systems. Modern technology using bar-code scanning and computer databases has enabled most merchandisers to adopt this system.

1.6. TRIAL BALANCE

In the ledger, the sum of all debit balances must always be equal to the sum of all credit balances. Accountants periodically test this equality by preparing a trial balance. Figure 1.13 is an example of a trial balance.

A trial balance reveals that whether or not the ledger is in balance, that is, equal debits and credits have been posted for all transactions in the ledger. When the sums of debit and credit columns of the trial balance are not equal, one or more of the following common errors may have occurred:

- Error in summation.
- Error in computing the balance of one or more accounts.
- Error in carrying the account balance to the trial balance.
- Error in posting a debit as a credit, or vice versa.
- Error in permuting digits of recorded amounts.

Periodic Inventory System

General Journal

Date	Description	Dr.	Cr.
2020			
May 3	**Purchases**	600	
	A/P – XYZ		600
	Bought 6 gadgets @ $100 (terms 5/10, n/30)		
5	A/P - XYZ	100	
	Purch. Returns & Allow.		100
	Return of 1 gadget to XYZ		
9	A/R - ABC	450	
	Sales		450
	Selling 3 Gadgets @ $150		
11	A/P – XYZ	500	
	Cash		475
	Purchases Discount		25
	Paid for 5 gadgets to XYZ		
15	**Sales Returns & Allowance**	150	
	A/R – ABC		150
	Return of 1 Gadget by ABC		

Perpetual Inventory System

General Journal

Date	Description	Dr.	Cr.
2020			
May 3	**Inventory**-Gadgets	600	
	A/P - XYZ		600
	Bought 6 gadgets @ $100 (terms 5/10, n/30)		
5	A/P - XYZ	100	
	Inventory-Gadgets		100
	Return of 1 gadget to XYZ		
9	A/R - ABC	450	
	Sales		450
	Selling 3 Gadgets @ $150		
11	A/P – XYZ	500	
	Cash		475
	Inventory – Gadgets		25
	Paid for 5 gadgets to XYZ		
11	**Cost of Goods Sold**	300	
	Inventory – Gadgets		300
	Cost of 3 Gadgets @ $100		
15	**Sales Returns & Allowance**	150	
	A/R – ABC		150
	Return of 1 Gadget by ABC		
15	**Inventory** – Gadgets	100	
	Cost of Goods Sold		100
	Adding return to inventory		

FIGURE 1.12 Comparing the general journal entries for purchases and sales transactions using the periodic and perpetual inventory systems.

However, there is no way the trial balance can detect the following types of errors:

- A debit (or a credit) of an account being posted improperly in another account.
- The omission of a transaction (not posted).
- An error of the same amount made in both debited and credited accounts.

1.7. CASH AND ACCRUAL BASES OF ACCOUNTING

Under the cash basis of accounting, expenses and revenues are accounted for whenever a cash payment is made or received, regardless of when the expenses accrued or the revenues earned.

Under the accrual basis of accounting, expenses and revenues are accounted for in the accounting period that they are realized regardless of when payment is made or cash is received.

The accrual basis follows the "matching rule" of accounting. Under this rule, revenues must be assigned to the accounting period in which the goods are sold or the services rendered, and expenses must also be assigned to the accounting period in which they are accrued.

XYZ Company
Trial Balance
As of July 31, 20XX

Account	Dr.	Cr.
Cash	$1,250	
Accounts Receivable	2,250	
Office Supplies	400	
Prepaid Insurance	360	
Office Equipment	22,540	
Accounts Payable		$1,700
Notes Payable		1,000
Mr. XYZ, capital		19,300
Mr. XYZ, withdrawals	2,000	
Fees Earned		8,000
Wages (Expense)	700	
Utility Expense	100	
Telephone Expense	50	
Rent Expense	350	
TOTAL	$30,000	$30,000

FIGURE 1.13 An example of a trial balance.

1.8. DEPRECIATION OF LONG-LIVED ASSETS

A firm buys long-lived assets (such as automobiles, machinery, equipment, and buildings) to use over a multiple number of years. Hence, the costs of such assets should be allocated to the years of the estimated useful lives. The annual allocated cost is called depreciation or depreciation cost/expense.

The main purpose of depreciation computation is to recover the capital invested in an asset whose value is normally declined as a result of time and usage.

Since depreciation expenses are matched against revenues, they reduce the income reported by the company and hence reduce income taxes. The cash preserved through tax reduction can, of course, be invested to earn more income in the next accounting period.

The Accumulated Depreciation account is a contra-asset account on a company's accounting records and balance sheet with a credit balance. It appears on the balance sheet as a reduction from the gross amount of the corresponding fixed asset. The remainder of this subtraction is the book value of the asset.

1.9. COMPLETING AN ACCOUNTING PERIOD

An accounting period ends on a specific date. The income statement must contain all revenues and expenses applicable to the period ended by this date, only. The balance sheet must contain all assets, liabilities, and owner's equity (on company's net worth) as of that date. For this purpose, accountants make adjusting entries to update account balances. Then, they use closing entries to close the accounting books and transfer the account balances to the books for the following fiscal year.

1.9.1. Adjusting Entries

Adjusting entries are made at the end of a fiscal period to the accounts for the transactions that have not been recorded. Accountants record the adjusting in the general journal and post them in the general ledger. The following are examples of items that may require adjusting entries:

- Unrecorded revenues earned from services rendered but not yet collected or billed to the customers.
- Unearned revenues that have been recorded as (earned) revenues.
- Accrued wages due to employees and the corresponding taxes that will be paid in the following fiscal period.
- Interest expense paid on loans and the loan balance updates.
- Accrued interest expense.
- Prepaid insurance and insurance expenses to account for the expired portions of prepaid insurance.
- Prepaid rent and rent expenses to account for the expired portions of prepaid rent.
- Supplies (inventory) and the supplies expense to account for the portions of supplies (inventory) used.
- Appraised inventories on hand at the period end.
- Accrued interest and dividend income from bank accounts and investments but not yet received.
- Depreciation expense and accumulated depreciation.
- Computed amount of bad debts and adjustment in accounts receivable.
- Accrued property taxes.
- Accrued income taxes.
- Updated net worth for net income or loss.
- Declared dividends (payable) due to stockholders.

After recording all adjusting entries in the general journal and posting them to the general ledger, the account balances are determined, and an "**adjusted trial balance**" is prepared. Using the account balances from the adjusted trial balance, accountants prepare and report the financial statements.

1.9.2. Closing Entries

Accountants use an "**income summary**" account to summarize all revenues and expenses, find the net income (or loss) and transfer the results to the Taxes Payable, Dividends Payable, and Capital (Retained Earnings) accounts. All transfers are also made by means of closing entries. Following are the required closing entries.

Required Closing Entries for a Single-Owner Company

- Transferring the balance of the revenue account to the Income Summary account.
- Transferring the balances of all expense accounts to the Income Summary account.
- Transferring the balance of the Income Summary account to the "Taxes Payable" and "Capital" accounts.
- Transferring the balances of the Withdrawals account to the Capital account.

Required Closing Entries for a Corporation

- Transferring the balances of revenue accounts to the Income Summary account.
- Transferring the balances of expense accounts to the Income Summary account.
- Transferring the credit balance of the Income Summary account to the "Taxes Payable," "Dividends Payable" (if declared to be paid), and "Retained Earnings" accounts. The Debit balance of Income Summary, which shows a net loss, is transferred to the Retained Earnings account only (because there is no profit to pay tax on or dividends to the stockholder from).

Example 1.3

Assume that at the end of year 20XX, a trial balance for McArden Dry Cleaning was prepared as shown in Figure 1.14. The following information became available after the trial balance was prepared:

a) A study of insurance policies shows that $170 is unexpired at the end of the year.
b) An inventory of cleaning supplies shows $414 on hand.
c) Estimated depreciation for the year was $4,300 on the building and $1,300 on the delivery truck.
d) Accrued interest on the mortgage payable amounted to $500.
e) On November 1, 20XX, the company signed a contract with Morgan County Hospital to dry clean, for a fixed monthly charge of $200, the uniforms used by doctors in surgery. The hospital paid for four months in advance.

McArden Dry Cleaning
Trial Balance
December 31, 20XX

Account	Debit	Credit
Cash	$ 1,256	
Accounts Receivable	10,280	
Prepaid Insurance	700	
Cleaning Supplies	2,687	
Land	9,000	
Building	75,000	
Accumulated Depreciation, Building		$ 14,200
Delivery Truck	11,500	
Accumulated Depreciation, Delivery Truck		2,600
Accounts Payable		10,200
Unearned Dry Cleaning Fees		800
Mortgage Payable		60,000
Ed McArden, Capital		23,642
Ed McArden, Withdrawals	12,000	
Dry Cleaning Revenue		57,200
Laundry Revenue		18,650
Plant Wages	32,560	
Sales and Delivery Wages	18,105	
Cleaning Equipment Rent	3,000	
Delivery Truck Expenses	2,187	
Interest Expense	5,500	
Other Expenses	3,517	
TOTAL	$187,292	$187,292

FIGURE 1.14 A trial balance for Example 1.3.

f) Unrecorded plant wages totaled $982.
g) Sales and delivery wages are paid on Friday. The weekly wages are $350. December 31, 20XX, was Thursday.
   - Record the adjusting entries in a general journal format.
   - Open T-accounts for each account in the trial balance plus the following: Wages Payable; Accrued Interest; Insurance Expense; Cleaning

Supplies Expense; Depreciation Expense, Building; Depreciation Expense, and Delivery Truck. Then, record the balances from the trial balance in the T-accounts. Post all adjusting entries also to the T-accounts.

- Prepare an adjusted trial balance.
- Prepare an income statement and balance sheet.

Solution

Adjusting entries are recorded in the general journal as shown in Figure 1.15. The T-accounts and posting balances and adjusting entries into them are shown in Figure 1.16. The adjusted trial balance is given in Figure 1.17. The income statement and balance sheet are given in Figures 1.18 and 1.19, respectively.

General Journal Page 914

DATE		DESCRIPTION	Post. Ref.	Debit	Credit
20XX					
Dec.	31	Insurance Expense	a	530	
		Prepaid Insurance	a		530
		(700 - 170 = 530)			
Dec.	31	Cleaning Supplies Expense	b	2,273	
		Cleaning Supplies	b		2,273
		(2687- 414 = 2,273)			
Dec.	31	Depreciation--Building	c	4,300	
		Depreciation--Truck	c	1,300	
		Accumulated Dep.--Build.	c		4,300
		Accumulated Dep.--Truck	c		1,300
Dec.	31	Interest Expense	d	500	
		Accrued Interest	d		500
Dec.	31	Unearned Dry Cleaning Fees	e	400	
		Dry Cleaning Revenue	e		400
		($200 x 2 = $400)			
Dec.	31	Plant Wages	f	982	
		Wages Payable	f		982
Dec.	31	Sale & Delivery Wages	g	280	
		Wages Payable	g		280
		([$350/5]×4 = $280 wages for last 6 days of Dec.)			

FIGURE 1.15 Recording the end-of-year adjusting entries for Example 1.3.

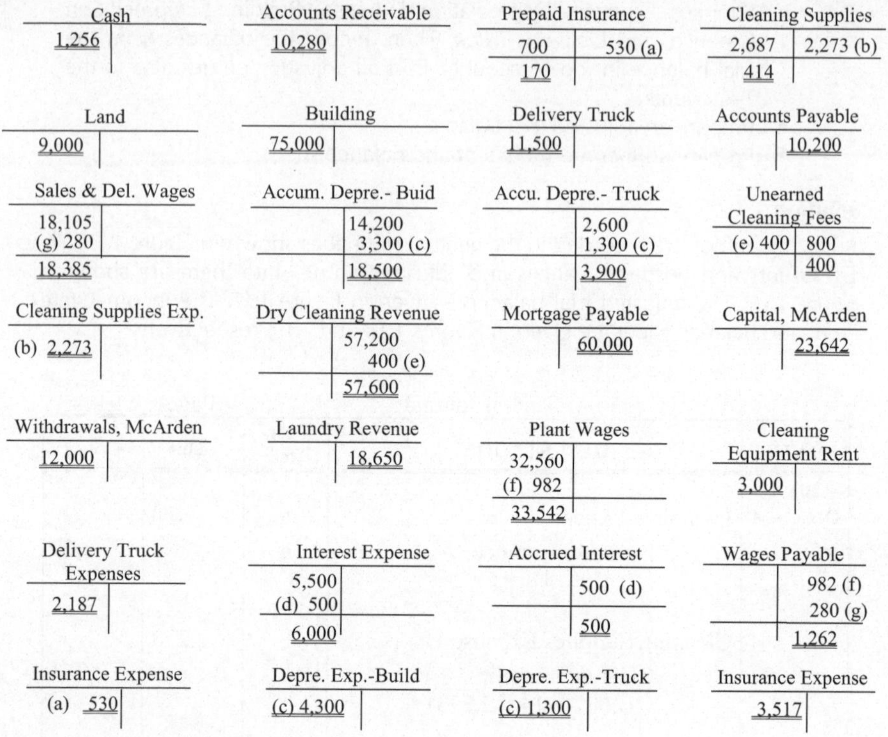

FIGURE 1.16 The T-accounts for Example 1.3.

1.10. TYPES OF ACCOUNTS

There are two major categories of accounts, which are listed below:

- **Balance Sheet Accounts**. These accounts (also called **permanent** or **real** accounts) usually carry an initial balance from the previous accounting period (or year), may increase or decrease during the current period, and carry an ending balance to the next accounting period. They include assets, liabilities, and capital accounts and are shown in the balance sheet.
- **Nominal Accounts**. This type of accounts includes all revenue- and purchases-related accounts and expenses accounts, which are temporary in nature. They begin each accounting period with a zero initial balance, accumulate a balance during the period, and their end-of-period balances are transferred to another temporary account, called Income Summary account, by means of closing entries.

McArden Dry Cleaning
Adjusted Trial Balance
As of December 31, 20XX

Account	Debit	Credit
Cash	$ 1,256	
Accounts Receivable	10,280	
Prepaid Insurance	170	
Cleaning Supplies	414	
Land	9,000	
Building	75,000	
Accumulated Depreciation, Building		$ 18,500
Delivery Truck	11,500	
Accumulated Depreciation, Del. Truck		3,900
Accounts Payable		10,200
Unearned Dry Cleaning Fees		400
Mortgage Payable		60,000
Capital		23,642
Withdrawals	12,000	
Dry Cleaning Revenue		57,600
Laundry Revenue		18,650
Plant Wages	33,542	
Sales and Delivery Wages	18,385	
Cleaning Equipment Rent	3,000	
Delivery Truck Expense	2,187	
Interest Expense	6,000	
Other Expenses	3,517	
Wages Payable		1,262
Accrued Interest Payable		500
Insurance Expense	530	
Cleaning Supplies Used	2,273	
Depreciation Expense, Building	4,300	
Depreciation Expense, Delivery Truck	1,300	
Totals	$194,654	$194,654

FIGURE 1.17 An adjusted trial balance for Example 1.3.

1.11. CLASSIFIED BALANCE SHEET AND INCOME STATEMENT

Balance sheets should be prepared in a classified fashion (Figure 1.20) to be informative and convenient for analysis. A common classification is presented as follows:

1.11.1. Division of a Balance Sheet

- **Assets**: A list of all resources owned by or owed to the company. The following are the three general classes of assets:

McArden Dry Cleaning **Income Statement** For the Period Ending December 31, 20XX		
Revenues:		
Dry Cleaning Revenue	$57,600	
Laundry Revenue	18,650	
Total Revenues		$76,250
Less: Operating Expenses:		
Plant Wages	$33,542	
Sales and Delivery Wages	18,385	
Cleaning Equipment Rent	3,000	
Delivery Truck Expense	2,187	
Insurance Expense	530	
Cleaning Supplies Used	2,273	
Depreciation Expense, Building	4,300	
Depreciation Expense, Delivery Truck	1,300	
Total Operating Expenses		65,517
Operating Income		$10,733
Less: Other Expenses:		
Interest Expense	$6,000	
Other Expenses	3,517	
Total Other Expenses		9,517
Net Income (before-tax)		$1,216

FIGURE 1.18 An income statement for Example 1.3. Notes: Since there is no information regarding the tax rate, it is assumed that the net income (before-tax) is not taxable and hence added to the capital as after-tax net income in the balance sheet.

- **Current Assets**: They are the company's working capital that can be converted into cash approximately within a year. They are listed in their order of faster cashability and include Cash, Marketable Securities, Accounts Receivable, Prepaid Expenses, Inventories, and Supplies.
- **Fixed Assets**: They are long-lived assets of the company. They include Land, Buildings, Trucks, Automobiles, Plant, and Equipment.
- **Intangible Assets**: These assets do not physically exist. They include Patent, Copyright, and Goodwill.
- **Liabilities**: A list of the company's financial obligations. The following are the two categories of liabilities:
- **Current Liabilities**: These are the company's short-term financial obligations that are due in the next accounting period. They include Accounts Payable, Notes Payable (due within a year), Wages Payable, Interest Payable, Accrued Expenses, Dividends Payable, and Taxes Payable.

McArden Dry Cleaning
Balance Sheet
As of December 31, 20XX

ASSETS			
Current Assets:			
Cash		$1,256	
Accounts Receivable		10,280	
Prepaid Insurance		170	
Cleaning Supplies		414	
Total Current Assets			$12,120
Fixed Assets:			
Land		$9,000	
Building	$75,000		
Less: Accum. Dep., Build.	18,500		
Book Value of Building		56,500	
Delivery Truck	$11,500		
Less: Accum. Dep., Truck	3,900		
Book Value of Deliver Truck		7,600	
Net Fixed Assets			73,100
TOTAL ASSETS			**$85,220**
LIABILITIES			
Current Liabilities:			
Accounts Payable		$10,200	
Unearned Dry Cleaning Fees		400	
Wages Payable		1,262	
Accrued Interest Payable		500	
Total Current Liabilities			**$12,362**
Long-Term Liabilities:			
Mortgage Payable			60,000
Total Liabilities			**$72,362**
OWNERS' EQUITY			
Capital--Beginning (1/1/20XX)		$23,642	
Add: Net Income (after-tax)	$ 1,216		
Less: Withdrawals	12,000		
Excess of Withdrawals over Net Income		(10,784)	
Owner's Equity: Capital—Ending (12/31/20XX)			12,858
TOTAL LIABILITIES AND OWNER'S EQUITY			**$85,220**

FIGURE 1.19 A balance sheet for Example 1.3.

- **Long-Term Liabilities**: These are the company's long-term financial obligations with a maturity date longer than a year. They include Notes Payable (due longer than one year), Mortgages Payable, and Bonds Payable.
- **Net Worth**: The interests of the owners in the company, which is also called the owner's (or stockholders') equity. They include Preferred Stock, Common Stock, and Retained Earnings.

McAlter Corporation
Balance Sheet
December 31, 20XX

ASSETS			
Current Assets:			
Cash		$36,758	
Marketable Securities		28,500	
Accounts Receivable		65,442	
Inventories		153,500	
Supplies		1,750	
Total Current Assets			$285,950
Fixed Assets:			
Land		$87,500	
Building	$684,000		
Less: Accumulated Depreciation	91,200		
Book Value of Building		592,800	
Equipment	$849,000		
Less: Accumulated Depreciation	195,800		
Book Value of Equipment		653,200	
Total Fixed Assets			1,333,500
Intangible Assets:			
Patents			120,000
TOTAL ASSETS (NET)			**$1,739,450**
LIABILITIES			
Current Liabilities:			
Accounts Payable		$ 65,950	
Wages Payable		3,250	
Interest Payable		3,750	
Dividends Payable		10,500	
Taxes Payable		18,675	
Total Current Liabilities			**$ 102,125**
Long-Term Liabilities:			
Notes Payable (due more than 1 year)		$ 56,000	
First Mortgage Bonds (8%)		181,000	
Total Long-Term Liabilities			**237,000**
Total Liabilities			**$ 339,125**
NET WORTH			
Preferred Stocks	$350,000		
Common Stocks (par value $20)	962,500		
Retained Earnings	87,825		
NET WORTH (Owner's Equity)		**1,400,325**	
TOTAL LIABILITIES AND NET WORTH			**$1,739,450**

FIGURE 1.20 An example of a classified balance sheet.

1.11.2. Formatted Income Statement

Figure 1.21 shows the common format of an income statement:

- **Sales Revenues**: The amounts received through sales of goods or services during a given fiscal period.
- **Cost of Goods Sold**: The net manufacturing cost of the product produced and sold by a company during a given fiscal period.
- **Gross Margin**: The difference in revenues and cost of goods sold.
- **Operating Expenses**: All expenses related to the company's normal operation incurred during the preceding accounting period.
- **Operating Income**: The amount of profit realized from a business's operations, after deducting operating expenses (i.e., cost of goods sold, selling and administrative expenses, and depreciation). It is equal to gross margin minus operating expenses.
- **Nonoperating Income**: The revenues received from sources other than sales of products or services, which include the interests and dividends earned from its investments, rent received through leasing its properties, and capital gains on selling assets.

McAlter Corporation
Income Statement
For the Year Ended December 31, 19XX

Revenues (Sales)		$1,747,500
Cost of Goods Sold		1,015,000
Gross Profit (from sales)		$ 782,500
Less: Operating Expenses:		
Selling Expenses	$99,500	
Administrative Expenses	122,000	
Other expenses (miscellaneous)	42,480	
Total Operating Expenses		263,980
Operating Income		$518,520
Other Income and Nonoperating Expenses:		
Add: Interest Revenue	$ 4,200	
Less: Interest Expense	36,620	
Less: Excess Expense over Income		32,420
Income Before Taxes		$486,100
Federal Income Taxes (50%)		243,050
Net Income (after-tax)		$243,050

FIGURE 1.21 An example of an income statement.

- **Nonoperating Expenses**: The expenses not related to the company's normal operation (e.g., interest expense).

SUMMARY

The accounting system maintains accurate and timely records of the daily financial transactions of a business. It compiles the transactions and prepares various important financial statements such as the income statement, owners' equity statement, balance sheet, and cash flow statement. Decision-makers utilize the financial statements for managerial and economic decisions.

REFERENCE

Committee on Accounting Terminology: *Accounting Terminology*, Bulletin No. 2., American Institute of Certified PublicAccountants, p. 9, New York, 1953.

REVIEW QUESTIONS AND ANSWERS

R1.1. What are revenues?
R1.2. What are an earned income and unearned income?
R1.3. What are operating and nonoperating revenues?
R1.4. What are prepaid expenses? Describe how accountants record and report them in accounting records.
R1.5. What is the difference between the operating income and the net income items?
R1.6. What are the items normally reported as current liabilities?
R1.7. What is the formula for the cost of goods sold?
R1.8. What is the cost of goods sold for merchandising and manufacturing firms?
R1.9. What is inventory?

PROBLEMS

P1.1. The following is the list of transactions for the John Smith Advertising Company during the first two weeks of February, 20XX:

1)	2/01	John Smith invested $25,000 in his advertising company.
2)	2/02	Leased an office and paid $2,000 for two months' rent in advance.
3)	2/02	Purchased some art equipment from Zoran Art Supplies on credit, $6,000.
4)	2/02	Hired an office assistant and agreed to pay her $500 every two weeks.
5)	2/03	Paid $300 cash for some office supplies.
6)	2/03	Paid $650 for a one-year insurance policy.
7)	2/04	Paid Zoran Art Supplies $1,000 of the amount owed.
8)	2/05	Performed an advertising service for a local store and delivered a fee invoice of $900 to the store. The store manager agreed to pay the invoice in two weeks.

9)	2/06	Received $500 in advance to perform an advertising service for a local car dealer by placing advertisements in the newspaper.
10)	2/12	John Smith withdrew $300 from the business for personal spending.
11)	2/13	Completed the service for the car dealer and sent them an invoice of $700 due after deducting the $500 advance receipt on 2/06 (Transaction #9).
12)	2/14	Paid the office assistant two weeks' salary, $500.

Required

a) Record the transactions in the company's general journal.

b) Open and name appropriate T-accounts and post the entries to them.

P1.2. The Bunyon Consulting Company is owned by William Woody. The following alphabetical list shows the account balances for the company as of June 30, 20XX:

Accounts Receivable	$206,281
Accounts Payable	145,895
Accumulated Depreciation, Building	171,200
Accumulated Depreciation, Equipment	162,702
Building	856,000
Cash	350,150
Depreciation Expense, Building	34,240
Depreciation Expense, Equipment	40,675
Equipment	406,750
Land	222,000
Marketable Securities	256,734
Mortgage Payable	545,000
Revenues	765,435
Salary Expense	210,653
Supplies	12,789
Supplies Expense	8,662
Utilities Expense	3,238
William Woody, Capital	892,940
William Woody, Withdrawals	?

Note: The "?" mark is a missing item to be determined.

Required

a) Prepare a trial balance for the Bunyon Consulting Company with the proper heading. Indicate the correct amount of William Woody, Withdrawals balance in the trial balance.

b) Prepare an "Income Statement" for the 6-month period ending June 30, 2005, for the company. Assume a 40% tax rate on the company's net income.

P1.3. At the end of the year 20XX, the trial balance for TBT Company, owned by Mr Ken Ram, appeared as follows:

TBT Company
Trial Balance
December 31, 20XX

Account	Dr.	Cr.
Cash	$21,742	
Accounts Receivable	15,986	
Prepaid Insurance	1,200	
Office Supplies	7,022	
Building	60,000	
Accumulated Depreciation, Building		$4,000
Equipment	25,000	
Accumulated Depreciation, Equipment		1,000
Accounts Payable		1,200
Mortgage Payable		40,000
Ken Ram, Capital		35,000
Ken Ram, Withdrawals	36,900	
Revenue		250,000
Telephone Expense	450	
Wages (Expense)	90,575	
Utilities Expense	12,325	
Administrative Expenses	40,000	
Interest Expense	20,000	
Total	$331,200	$331,200

The following information was also available:

1) Supplies on hand, on December 31, 20XX, were $2,232.
2) Insurance still unexpired amounted to $450.
3) Estimated depreciation of equipment was $1,000.
4) Estimated depreciation of building was $2,000.
5) A telephone bill has been received on December 31 showing $100 telephone usage during December but has not been recorded yet.

Required

a) Record the necessary adjusting entries.
b) Open T-accounts for the accounts given in the trial balance plus the following: Office Supplies Expense; Insurance Expense; Depreciation Expense, Equipment; Depreciation Expense, Building. Record the balances shown in the trial balance in the T-accounts.
c) Post the adjusting entries to the T-accounts.
d) Prepare an adjusted trial balance.
e) Prepare an income statement (assume an income tax rate of 40%).
f) Prepare a balance sheet.

P1.4. The following accounting information at the end of year 20XX is available for Satoosh Corporation:

Account	Balance
Cash	$125,500
Accounts Receivable	115,000
Accounts Payable	121,500
Mortgage Payable	120,000
Common Stocks Outstanding	90,000
Retained Earning	30,000
Equipment	125,000
Building	135,000
Accumulated Depreciation, Equipment	37,500
Accumulated Depreciation, Building	18,000
Prepaid Insurance	1,200
Office Supplies	2,000
Office Supplies Used (Expense)	7,000
Revenue	400,000
Selling and Administrative Expenses	232,800
Power and Utilities Expense	38,500
Interest Expense	18,000
Depreciation Expense, Equipment	12,500
Depreciation Expense, Building	4,500

Assume that all end-of-year adjusting entries have already been recorded; that is, the accounts listed above do not need any adjustment.

Required

a) Prepare an income statement (assume income tax rate of 40%).

b) Prepare a classified balance sheet.

P1.5. Three construction engineers established TADJ Construction Corporation in early 20XX. The company had completed the following transactions in the month of February 20XX:

Feb. 1. The founders contributed the following assets to the company and received 2,000 shares of preferred stocks of $100 par value:

Cash	$30,000
Office Equipment	20,000
Land	150,000

Feb. 2. A total of 80,000 shares of common stocks of $10 par value were issued and sold.

Feb. 4. A building was purchased for $500,000, of which $50,000 was paid as down payment and the rest was by a 12% mortgage.

Feb. 5. Paid $24,000 for an insurance policy for the next 12 months.

Feb. 6. Paid $1,000 for advertisement announcing the establishing of the company and its grand opening.

Feb. 7. Purchased construction equipment costing $500,000. Paid $50,000 cash and a note was issued for the balance.

Feb. 10. Received a bill of commissions charged by the broker who sold the stocks, $16,000.

The following accounts are established by the company's accountants:

Cash, Prepaid Insurance, Land, Building, Office Equipment, Construction Equipment, Accounts Payable, Notes Payable, Mortgages, Common Stocks, Preferred Stocks, Advertising Expense, Commission Expense.

Required

a) Prepare journal entries to record the above transactions in the general journal.

b) Post the entries of part (a) to the related T-accounts.

P1.6. The following trial balance was available for Roofing Company at the end of the year 20XX.

Roofing Company
Trial Balance
As of December 31, 20XX

Account	Dr.	Cr.
Cash	$160,000	
Accounts Receivable	220,000	
Office Supplies	20,000	
Inventories	250,000	
Equipment	300,000	
Accumulated Depreciation, Equipment		$60,000
Building	500,000	
Accumulated Depreciation, Building		100,000
Accounts Payable		200,000
Mortgage Payable		400,000
Common Stocks		300,000
Retained Earnings		150,000
Revenue (Sales)		930,000
Cost of Goods Sold	400,000	
Selling Expenses	140,000	
Administrative Expenses	110,000	
Interest Expense	40,000	
Totals	$2,140,000	$2,140,000

Required

a) Prepare an income statement (assuming 50% tax rate).

b) Prepare a classified balance sheet (assume that the company declares 40% of net income as dividends).

APPENDIX 1.A

A Chart of Accounts Sample

Typical accounts found in the chart of accounts are (code numbers are suggested by the author):

No.	Account Name	Type of Accounts
1XX	**ASSETS**	
11X	**Current Assets:**	
111	Cash (checking account)	Current Asset
112	Petty Cash	Current Asset
113	Marketable Securities	Current Asset
114	Accounts Receivable	Current Asset
115	Allowance for Doubtful Accounts	Current Asset (contra)
106	Prepaid Expenses	Current Asset
107	Inventory	Current Asset
12X	**Fixed Assets:**	
121	Land	Fixed Asset
122	Building	Fixed Asset
123	Accumulated Depreciation – Building	Fixed Asset (contra)
124	Equipment	Fixed Asset
125	Accumulated Depreciation – Equipment	Fixed Asset (contra)
126	Machineries	Fixed Asset
127	Accumulated Depreciation – Machineries	Fixed Asset (contra)
13X	**Intangible Assets:**	
131	Patents	Intangible Asset
131	Copyrights	Intangible Asset
132	Goodwill	Intangible Asset
2XX	**LIABILITIES**	
21X	**Current Liabilities:**	
211	Accounts Payable	Current Liabilities
212	Mortgage Payable (due within a year)	Current Liabilities
213	Bonds Payable (due within a year)	Current Liabilities
214	Unearned Income	Current Liabilities
215	Accrued Expenses	Current Liabilities
216	Wages Payable	Current Liabilities
217	Taxes Payable	Current Liabilities
218	Dividends Payable	Current Liabilities
22X	**Long-Term Liabilities:**	
221	Notes Payable (due longer than a year)	Long-Term Liabilities
222	Mortgage (due longer than a year)	Long-Term Liabilities
223	Pension Liabilities	Long-Term Liabilities
3XX	**STOCKHOLDERS' EQUITY**	
311	Common Stock	Net Worth

No.	Account Name	Type of Accounts
312	Preferred Stock	Net Worth
313	Retained Earnings	Net Worth
4XX	**REVENUES**	
411	Revenue (or Sales)	Revenue
412	Sales Returns and Allowances	Revenue (contra)
413	Sales Discounts	Revenue (contra)
421	Service Revenues	Revenue
422	Fees Earned	Revenue
414	Sales Taxes	Revenue (contra)
431	Interest Income	Revenue
5XX	**PURCHASES & EXPENSES**	
501	Purchases	Purchase
502	Purchases Returns and Allowances	Purchases (contra)
503	Purchases Discounts	Purchases (contra)
51X	**Cost of Sales:**	
511	Cost of Goods Sold (or Cost of Sales)	Expense
52X	**Selling Expenses:**	
521	Advertising and Promotions Expenses	Expense
522	Sales Salaries and Wages	Expense
523	Sales Force Payroll Taxes	Expense
524	Delivery Expense (Freight-Out)	Expense
525	Cleaning Expense – Sales Department	Expense
526	Depreciation Expenses – Sales Building	Expense
527	Accumulated Depreciation– Sales Building	Expense (contra)
528	Depreciation Expenses – Sales Vehicles	Expense
529	Accumulated Depreciation– Sales Vehicles	Expense (contra)
53X-55X	**Administrative Expenses:**	
531	Administrative Salaries Expense	Expense
532	Administrative Other Compensations	Expense
533	Administrative Payroll Taxes	Expense
541	Insurance Expense – Administrative Building	Expense
542	Maintenance Expense – Headquarter	Expense
543	Repair Expense – Headquarter	Expense
544	Supplies Expense – Headquarter	Expense
545	Utilities Expense – Headquarter	Expense
546	Depreciation Expense – Admin. Building	Expense
547	Accumulated Depreciation – Admin. Building	Expense (contra)
548	Depreciation Expenses – Admin. Vehicles	Expense
549	Accumulated Depreciation – Admin. Vehicles	Expense (contra)
551	Sundry (Other) Expenses	Expense
56X	**Nonoperating Expenses:**	
561	Bank Fees	Expense (nonoperating)
562	Interest Expense	Expense (nonoperating)
591	**Income Tax Expense**	Expense

2 Cost Analysis Fundamentals

LEARNING OBJECTIVES

- Learn the manufacturing costs.
- Define and distinguish among various types of costs.
- Learn basic cost formulation.
- Compute manufacturing unit cost.
- Compute average, marginal, incremental, and total costs.

2.1. INTRODUCTION

Costs are the necessary expenditures that a business must spend to produce products or provide services. Every factor of production has its associated costs, such as material, labor, and overhead costs.

To remain competitive in a market, the management needs to know whether or not certain products yield satisfactory profits. A product can successfully capture the market if its "*price is right*!" Therefore, knowledge of the unit cost of each product is necessary for setting its sales price. Higher manufacturing costs are recovered by setting higher sales prices. However, market price–demand relation imposes a ceiling on the sales price.

Deciding a sales price for a product is a critical managerial function. Before deciding on the price tag of a product, it becomes essential to obtain a close (ideally, exact) cost assessment of it. There are two types of inadequate product costing with undesirable economic consequences. On the one hand, if the unit cost of a product is over-estimated, there is a risk of over-pricing the product, which makes it highly likely to lose part or all of its market share to the competitors. On the other hand, if the unit cost of the product is under-estimated, there is a high chance of setting its sales price high enough that may result in losses.

Product costing and price-setting decisions are particularly complex for manufacturers who produce multiple products. For such manufacturers, information concerning product costing (i.e., assigning manufacturing costs) for each product becomes the key factor for successful marketing. This chapter presents the fundamental concepts and procedures of cost analysis. The techniques for cost allocation to multiple products will be presented in Chapter 5, "Joint Cost Allocation."

DOI: 10.1201/9780429432163-2

2.2. COSTS AND EXPENSES

Costs are expenditures for acquiring assets, while expenses refer to the consumption of the items acquired. A cost would become an expense if it has expired or was necessary to exchange to earn revenues and no longer belongs to the firm.

An example of costs is a firm's property insurance premium paid for the next 12 months of insurance. This advance payment is initially recorded as the current asset (in an account titled Prepaid Insurance, or Prepaid Expenses) since the cost has not been used up (has not expired) if the firm cancels the insurance, it can get a refund for the unused (unexpired) part of insurance. Otherwise, as time goes by, the firm reduces this cost (asset value) and converts it to insurance expense. Another example of costs is the manufacturing costs that a firm incurs for producing its products. These costs remain as current assets in the firm (as Materials, Work-In-Process, and Finished Goods Inventories). When the firm sells its manufactured items, it converts the Finished Goods Inventory into an expense as "*Cost of Goods Sold*" or "*Cost of Sales.*"

2.3. PRODUCT COSTING

Product or production costs refer to the costs that are incurred for manufacturing a product or providing a service. Manufacturing costs of a product include costs of direct labor, indirect labor (e.g., supervisory, human resources staff, maintenance, warehousing, security, and custodial services), direct (or raw) materials, indirect materials (consumable manufacturing supplies), and general factory overheads (e.g., rent, insurance, and property taxes). Costs of freight-in (shipping and handling for receiving materials) are added to the cost of materials, whereas costs freight-out (i.e., for delivering the product) to a customer are usually treated as periodic sales expenses.

Whenever the sales price of a product falls below its production cost, producers should try cost cutting if possible. Otherwise, they may have to lower or even shut down the production, temporarily or permanently. For example, the average price of oil was over $90 per barrel during 2011–14; it fluctuated between in the range of $30 and $75 during 2014–19. It collapsed (to around $12 per barrel) in April 2020 due to the COVID-19 pandemic effect on economic slowdown and demand for oil. The Organization of Petroleum Exporting Countries (OPEC) agreed to historic production reduction to stabilize prices. If the production cost of oil is over $20 per barrel, then the producers face significant negative cash flows (losses).

The product cost is often referred to as "*inventoriable cost.*" Companies record the costs of their unsold products as assets (*Finished Goods Inventory*) on their accounting records and report them on the balance sheet. When they sell their products, they record the costs of units sold as an expense (*Cost of Goods Sold*) and deduct the costs from their sales revenues to determine and report their profits (losses) on the income statement.

2.4. CLASSIFICATION OF COSTS

An engineering economic and cost analysis deals with various types of costs. To be able to perform such analysis, it is necessary to understand cost terminologies and classifications. This section presents various types of costs, based on their uses or purposes, with a brief description of each.

2.4.1. Classification of Costs by Behavior

The manufacturing (direct and indirect) cost is the total cost of all resources consumed in the process of making the product or providing the service. Cost behavior refers to the relationship between total **costs** and activity level. Based on the behavior, there are three categories of costs: fixed, variable, or semi-variable (or semi-fixed) costs. The fixed are further subcategorized based on flexibility and controllability. A summary of these classifications is as follows:

- Variable costs
- Fixed costs
  - Classification of fixed cost by flexibility
    - Committed fixed costs.
    - Discretionary fixed costs.
  - Classification of fixed cost by controllability
    - Avoidable (evitable or controllable) fixed costs.
    - Unavoidable (inevitable or uncontrollable) fixed costs.
- Semi-variable (semi-fixed) costs
  - Mixed (variable and fixed) costs.
  - Step-variable fixed costs.

Variable Costs: A variable cost (in total amount) varies in direct proportion to the changes in the level of activity (i.e., number of units produced or amount of service provided). When more units of a product are produced (or an activity level increases), the total variable cost increases by a multiplier. The costs of direct materials and labor are usually variable because they change in direct proportion to the production output. However, the variable cost per unit remains constant regardless of the production quantity. The total variable cost is the product of the variable cost per unit and the number of units produced, as expressed by the following formula and illustrated in Figure 2.1:

$$\mathbf{TVC} = f(Q) = V \cdot Q \tag{2.1}$$

where

TVC = Total variable cost;
V = Unit variable cost (dollars per unit);
Q = Output level (quantity or volume);
$f(Q)$ = A function of Q.

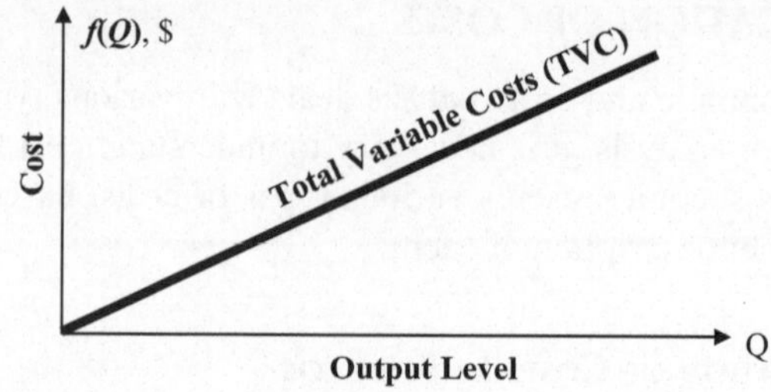

FIGURE 2.1 An illustration of the *total variable cost* as a function of output level.

Examples of variable costs include direct (raw) materials, direct labor, packaging costs, royalties, and commission as a percentage of the price.

Fixed Costs: A fixed cost (in total amount) remains constant regardless of the change in the level of production output, within the relevant range. *Relevant range* is the volume of output, over which cost behavior stays valid. However, the fixed cost per unit varies with the number of units; that is, as the level of activity increases, its fixed cost per unit decreases, as the fixed cost per unit (**FPU**) equals the total fixed cost (***F***) divided by the number of units produced (***Q***), as expressed by the following formula, and graphically illustrated in Figure 2.2.

$$\mathbf{FPU} = \boldsymbol{F/Q} \tag{2.2}$$

Examples of fixed costs include supervisory salaries, advertising, cost of hazard insurance, warehouse, depreciation on equipment, rent and property taxes, maintenance services, human resources, first-aid and medical services, and safety and security regardless of the level of activities.

Semi-Variable Costs: Semi-variable costs (or *semi-fixed*) are combinations of fixed and variable costs. These costs vary (change) with output but not in direct proportion. The fixed cost element is the part of the cost that is incurred regardless of output or activity level. The variable component of the cost varies in proportion to the level of output or activity. For this reason, these costs are also called *mixed costs.*

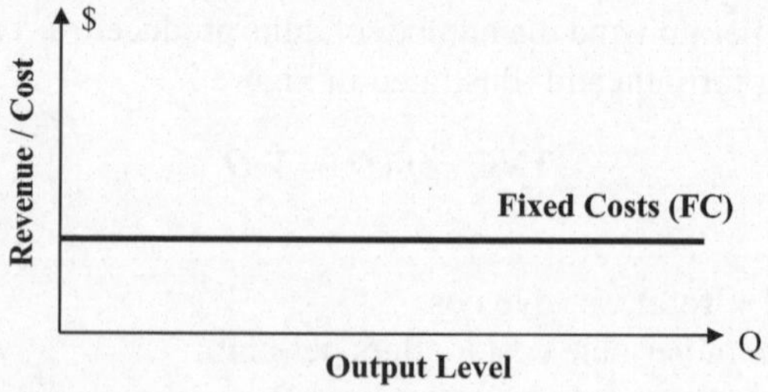

FIGURE 2.2 An illustration of the relationship between *fixed costs* and output level.

Common examples of semi-variable costs are utility and phone bills. For example, most bills for water, gas, and electricity will consist of a fixed subscription charge and a variable charge, which varies with the usage. Telephone bills, also, may contain a fixed line charge and call charges vary with the number of calls, the distance of the call, and the time duration. Figure 2.3 illustrates the behavior of mixed costs.

There are, also, cases in which the cost is **step-fixed**; that is, the cost does not change within certain ranges of activity, but it will change when the activity falls out of each range. In other words, the step-fixed costs vary step-wise, as illustrated in Figure 2.4.

Total Costs: Total cost of a product or service is the sum of all variable costs, fixed costs, and semi-variable (mixed) costs for its given output level. However, the total cost is usually the sum of total fixed costs and total variable costs, as expressed in the following formula and graphically illustrated in Figure 2.5.

$$\mathbf{TC} = \boldsymbol{V} \cdot \boldsymbol{Q} + \boldsymbol{F} \tag{2.3}$$

where

TC = Total cost (may be designated by C);
$V \cdot Q$ = Total variable cost (may be designated by VC);
V = Unit variable cost (dollars per unit);
Q = Output level (quantity or volume);
F = Total fixed cost (may be designated by TF).

To develop a cost equation, a company must first determine the amount of expense associated with producing and selling each of its products.

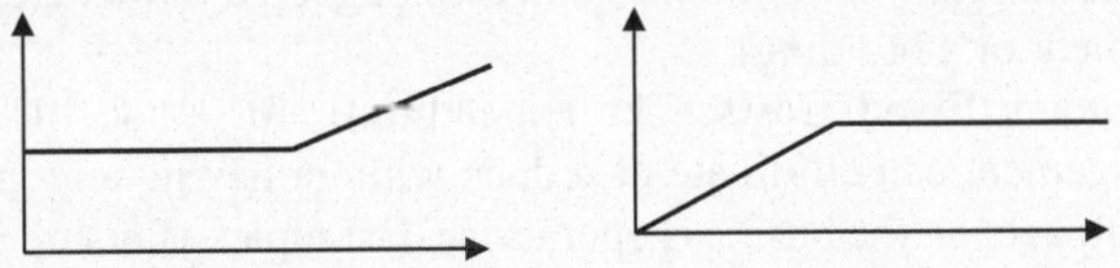

FIGURE 2.3 An illustration of a *mixed cost* as a function of output level.

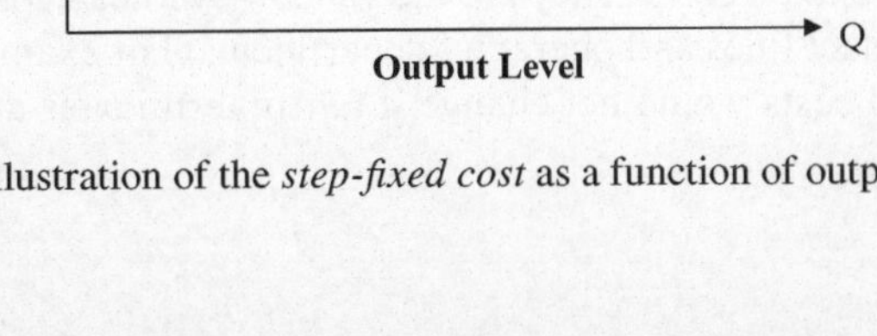

FIGURE 2.4 An illustration of the *step-fixed cost* as a function of output level.

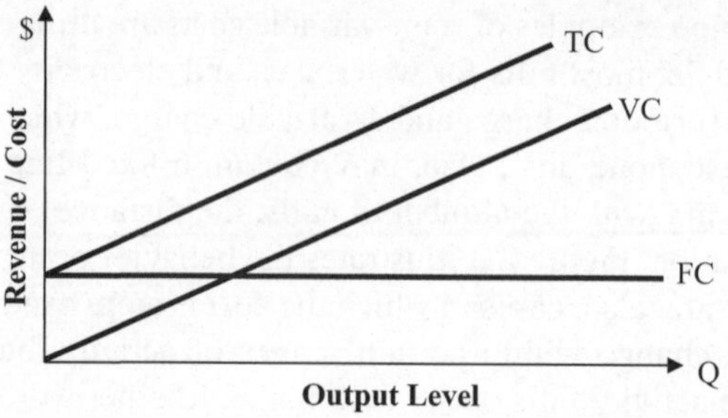

FIGURE 2.5 A graphical illustration of the *fixed* **(FC)**, *variable* (VC), and *total costs* (TC) against output level.

It must also determine the overhead costs associated with production. The cost equation is typically the cost of manufacturing and selling one item multiplied by the number of items sold and added to the company's overhead costs. For example, a company with $300,000 in overhead costs that makes products that cost the company $5 each to manufacture and sell has a cost equation of 300,000 + (5 × number of products sold).

The fixed costs are further subcategorized by "flexibility" and by "controllability" as described in the following subsections.

2.4.1.1. Classification of Fixed Costs by Flexibility

- **Committed Fixed Costs:** The expenditures that an organization has already made and cannot change them, and the financial obligations that the firm has already committed to and cannot get out of them (e.g., the lease of equipment or a building).
- **Discretionary Fixed Costs:** The expenditures for a specific period that the management can eliminate or reduce without having any immediate or serious impacts on the business operations. Examples of discretionary fixed costs are the costs of advertising, employee training, public relations, and research and development.

2.4.1.2. Classification of Fixed Costs by Controllability

- **Avoidable (Controllable) Fixed Costs**: A decision-maker's action can influence a controllable (or avoidable) fixed cost. For example, the fixed costs that are directly associated with the operation of a product line or department can be eliminated or reduced if the production line is shut down.
- **Unavoidable (Uncontrollable) Fixed Costs**: Unavoidable costs are unchangeable fixed costs. They are the general overhead costs that are allocated to product lines and operating departments. For example, the general factory fixed costs would not change if a firm terminates a product line or department.

2.4.2. Classification of Costs by Traceability

Cost traceability refers to whether costs can conveniently and economically traceable to specific products or departments. Based on traceability, there are two categories of costs, direct and indirect.

Direct Costs: A direct cost is any cost that is conveniently and economically traceable to a specific product manufactured or a service provided. The direct costs are more of concern for a firm that produces various types of products. Examples of this type of cost are direct materials and direct labor, which usually vary in proportion with the output level. Thus, direct costs are also variable costs.

Indirect Costs: An indirect cost is any cost that is not directly related (or conveniently and economically traceable) to any specific product or department. They are the general costs incurred to support all production lines or operation departments but not directly for any specific product or department. This type of costs includes all the overhead costs that are allocated to all products or operation departments.

For example, assume that a furniture manufacturer makes chairs and tables. The chairs produced use a certain type of wood, but the tables use another type of wood. In such a case, the types of wood would be direct costs, because each type of wood is directly related to a particular product. The wages paid to workers who work on both types of products, the depreciation, and the cost of power used to run machinery are indirect.

Prime Costs and Conversion Costs

A firm's management may need to make decisions based on the costs categorized as prime costs or conversion costs (Figure 2.6).

- **Prime Cost**: The prime cost is the sum of direct material costs, direct labor costs, and all other direct costs. A firm should calculate the prime cost for each of the product it manufactures to ensure they generate profit.

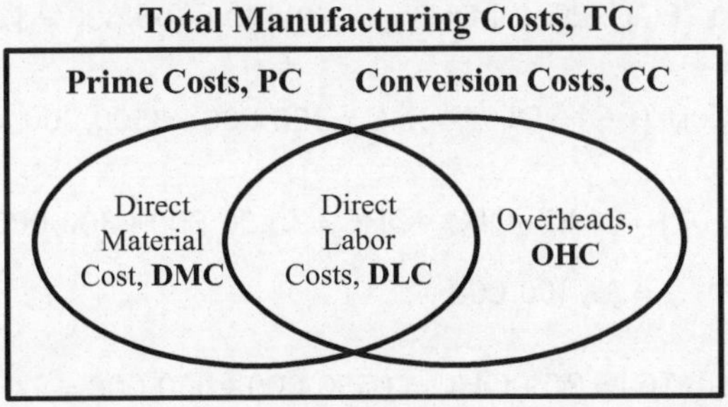

FIGURE 2.6 Prime costs and conversion costs. Prime costs and conversion costs do not sum up to total costs as direct labor is included in both.

- **Conversion Cost**: The conversion cost is the sum of direct labor and factory overhead costs. A firm incurs this cost to convert the raw materials into finished products.

Note that the total manufacturing cost (TC) equals direct material cost (DMC) plus direct labor cost (DLC) plus overhead cost (OHC); then,

$$\textbf{Prime Costs}\ (\textbf{PC}) = \textbf{DMC} + \textbf{DLC} \tag{2.4}$$

$$\textbf{Conversion Costs}\ (\textbf{CC}) = \textbf{DLC} + \textbf{OHC} \tag{2.5}$$

$$\textbf{Total Cost}\ (\textbf{TC}) = \textbf{DMC} + \textbf{DLC} + \textbf{OHC} \tag{2.6}$$

$$\textbf{Total Cost}\ (\textbf{TC}) = \textbf{PC} + \textbf{OHC} \tag{2.7}$$

$$\textbf{Total Cost}\ (\textbf{TC}) = \textbf{DMC} + \textbf{CC} \tag{2.8}$$

Example 2.1 (Manufacturing prime, conversion, and total costs)

A firm incurred the following costs in a year to manufacture 500,000 units of its single product:

Direct material costs: $1,250,000
Direct wages for workers: $300,000
Overheads (rent and general expenses): $850,000

Compute the prime, conversion, and total costs using equations (2.4) through (2.8) as they may apply.

Solution

$$\text{Prime Cost (PC)} = \text{DMC} + \text{DLC} = 1{,}250{,}000 + 300{,}000 = \$1{,}550{,}000$$

$$\text{Conversion Cost (CC)} = \text{DLC} + \text{OHC} = 300{,}000 + 8500{,}000 = \$1{,}150{,}000$$

$$\text{Total Cost (TC)} = \text{DMC} + \text{DLC} + \text{OHC} = 1{,}250{,}000 + 300{,}000 + 850{,}000$$
$$= \$2{,}400{,}000$$

$$\text{Total Cost (TC)} = \text{PC} + \text{OHC} = 1{,}550{,}000 + 850{,}000 = \$2{,}400{,}000$$

$$\text{Total Cost (TC)} = \text{DMC} + \text{CC} = 1{,}250{,}000 + 1{,}150{,}000 = \$2{,}400{,}000$$

2.4.3. Cost Classification by Decision-Making

There are various types of costs that are useful for business decision-making. They include the average cost per unit (or unit cost), marginal costs, incremental costs, opportunity costs, and sunk costs.

Average (Unit) Cost: An average cost (or unit cost) is the cost of each unit of a product, which equals the total cost of the product divided by the number of units produced. The following formula gives the average cost:

$$\textbf{Average (Unit) Cost} = \frac{\textbf{Total Cost}}{\textbf{Total Number of Units Produced}} \tag{2.9}$$

or

$$\mathbf{AC} = \frac{V \cdot Q + F}{Q} = V + \frac{F}{Q} \tag{2.10}$$

where

Q = Production quantity (units);
F = Total fixed cost;
V = Variable cost per unit;
VQ = Total variable cost;
AC = Average cost (or unit cost, UC).

However, this is only possible if the manufacturer produces just one type of product. If it produces a number of different products, it would be necessary to allocate the material, labor, and overhead costs appropriately to products to arrive at the total cost for each.

Example 2.2 (Unit cost)

A firm incurred the following costs in a year to manufacture 500,000 units of its single product:

Direct material costs: $1,250,000
Direct wages for workers: $300,000
Overheads (rent and general expenses): $850,000

The firm's total product cost (TC) is:

$$\text{TC} = \$1,250,000 + \$300,000 + \$850,000 = \$2,400,000$$

The product unit cost (UC) is:

$$\text{UC} = \$2,400,000/500,000 \text{ units} = \$4.80 \text{ per unit}$$

The unit cost is a basis for measuring profitability. For example, if the unit cost, including fixed and variable expenses, as in the previous example, is $4.80 per unit and the firm sells the product at $6.10 per unit, it generates a profit of $1.30 for each unit sold. A selling price of $4.00 creates a $0.80 loss per unit sold. However, this analysis does not necessarily capture all market activities. For instance, there may be additional costs due to sales returns shipping and disposals.

Marginal Cost: A marginal cost is an additional cost associated with producing an additional unit of the product or activity. The marginal cost equals the variable cost per unit if the production is within the relevant range. The marginal labor cost is the amount of time taken to produce an additional unit of the product multiplied by the time rate of labor cost.

Example 2.3 (Marginal cost)

A firm incurred the following costs in a year to manufacture 500,000 units of its single product:

Direct material costs: $1,250,000
Direct labor cost: $300,000
Overheads (rent and general expenses): $850,000

The firm's marginal cost (MC) is:

$$\text{MC} = (\$1,250,000 + \$300,000)/500,000 = \$3.10$$

Incremental Cost: An incremental cost (also known as a *differential* cost) is the change in total costs due to a change in a business activity or switching from one system (alternative) to another with a higher cost. One can determine the incremental cost by finding the additional costs involved in producing an additional unit of a product or adding another staff or machine, or adding a second shift. The change in the revenues of two alternatives is termed *incremental revenue.* Reviewing the incremental costs can help companies improve their operations' efficiency and profitability. The following is the formula for computing the incremental cost per unit:

$$\textbf{Incremental Cost per Unit} = \frac{\textbf{Change in Total Cost}}{\textbf{Number of Additional Units Produced}} \tag{2.11}$$

An incremental/differential cost analysis is conducted when making decisions such as: make or buy an item, change in the level of activity (product produced or service provided), add or eliminate a product line or department, change in product mix, and marketing goods in a new market. In an analysis, the analyst only considers variable costs as the "*relevant costs.*" Fixed costs and sunk costs, which would not change by a new action, are "*irrelevant costs.*" Future costs, which are mainly variable, are relevant to consider. The use of differential costs is only for making management decisions and has no accounting recordings.

Example 2.4 (Incremental cost)

Consider a machine shop that manufactures a certain product. The total fixed cost is $300 per day, which does not change regardless of the number of employees or the level of production. The total variable cost varies by the rate of $100 per employee per day, which in total is equal to the number of employees hired multiplied by this wage rate. For example, if three employees are hired, then the total variable cost equals 3 × $100 = $300. The number of units (quantity) of the product produced corresponding to the number of employees (labor) is as follows.

Labor	Quantity
1	20
2	45
3	65
4	80
5	88
6	93
7	93

Determine the total cost, unit cost, and incremental cost per unit corresponding to the number of employees hired.

Solution

Table 2.1 shows the solution in column 5 (total cost), in column 6 (unit cost), and in column 7 (incremental cost per unit).

- The total cost (column 5) is the sum of the fixed cost (column 3) and the total variable cost (column 4). For example, with four employees, the output will be 80 units and the total cost is $300 + $400 = $700.

TABLE 2.1
Output and Total Costs Summary

(1) Labor	(2) Production Quantity	(3) Fixed Cost	(4) Variable Cost	(5) Total Cost	(6) Unit Cost	(7) Incremental Cost/Unit
0	0	$300	$0	$300	—	—
1	20	300	100	400	$20.00	$5.00
2	45	300	200	500	11.11	4.00
3	65	300	300	600	9.23	5.00
4	80	300	400	700	8.75	6.67
5	88	300	500	800	9.09	12.50
6	93	300	600	900	9.68	20.00
7	93	300	700	1,000	10.75	∞

- The unit cost or average cost (column 6) is equal to the total cost (in column 5) divided by the output quantity (column 2). For instance, with four employees, for which the output is 80 units and the total cost is $700, the unit cost is $700/80 = $8.75 (rounded to the nearest cents).
- The incremental cost per unit (column 7) for each increment in production equals the change in the total cost divided by the change in the output quantity. For instance, with four employees, the change in the total cost is $700 – $600 = $100, the change in the output quantity is 80 – 65 = 15 units, and the incremental cost per unit is $100/15 = $6.67.

Figure 2.7 graphically illustrates how output affects the total cost. At zero (no) production, the total fixed cost is still $300. The fixed cost is the vertical axis intercept of the total cost curve. The graph shows that as production is increased, the variable costs (in total) and, consequently, the total cost would also increase.

Figure 2.8 illustrates the relationship between the output and the unit cost, and the incremental cost per unit. As production increases, unit cost decreases from an output level of 20 to 80 units; then it is gradually rising. This explains why the total cost increases at a decreasing rate up to an output level of 80 units and then increases at an increasing rate. This is due to the fact that the unit cost is the rate of change in the total cost.

The increase in the unit cost is caused by the economic ***law of diminishing returns*** (or the principle of diminishing marginal productivity), which states as an input for producing a commodity increases, while all other inputs are held fixed, at some point at which addition of that input generates progressively smaller, or diminishing, increases in output (Encyclopædia Britannica, 2020). As the number of employees increases from zero to one, output increases from 0 to 20 units for a marginal gain of 20 units; as the number rises from one to two employees, output increases from 20 to 45, a marginal gain of 25 units in the output. Beyond

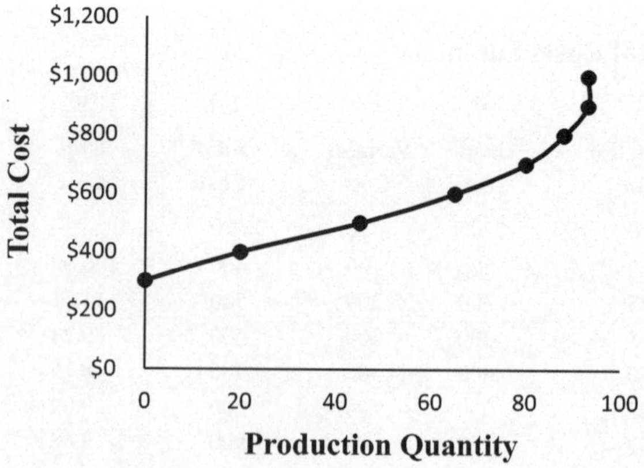

FIGURE 2.7 An illustration of how output changes affect the total cost for Example 2.4.

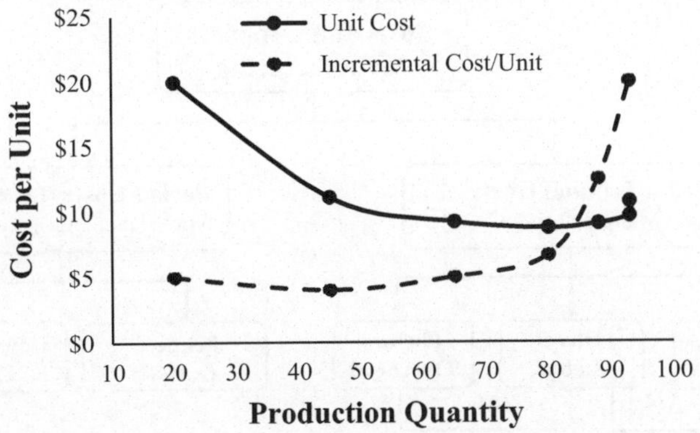

FIGURE 2.8 Relationship between the output and the unit (average) cost, and the incremental unit cost for Example 2.4.

that point, the marginal gain in output diminishes as each additional employee is hired.

The reason for the incremental cost per unit becomes lower than the current unit cost is the fixed costs may remain unchanged or minimally increase. Although the fixed costs may increase as production increases, the cost per unit declines if the company uses the current facilities and does not have to acquire additional equipment or facilities.

Opportunity Costs: An opportunity cost (or historical cost) is the profit lost when choosing an alternative instead of another. The concept is useful simply as a reminder to evaluate and compare all feasible alternatives before making the final decision. For example, a firm invests $1,000,000 and opens a new line of product that will have an annual return of 8%, or $80,000. If the firm could invest the fund in the expansion of its current product that would have generated an additional annual profit of $95,000, or a return of 9.5%, then the 1.5%, or $15,000, difference between the two alternatives is the foregone opportunity cost of the decision made.

Sunk Costs: A sunk cost is a cost that a firm incurred in the past, which it cannot recover. Sunk costs should not be considered when making the decision about investing in a new project, since the sunk costs are irrelevant, as they cannot be recovered. Instead, the only relevant cost of a current (old) project (e.g., a production machine) in replacement analysis is its market value if continued.

2.4.4. Classification of Costs by External Reporting

A manufacturing enterprise reports two types of costs: product (also known as *inventoriable* and manufacturing) costs and period (expenses) costs (Figure 2.9).

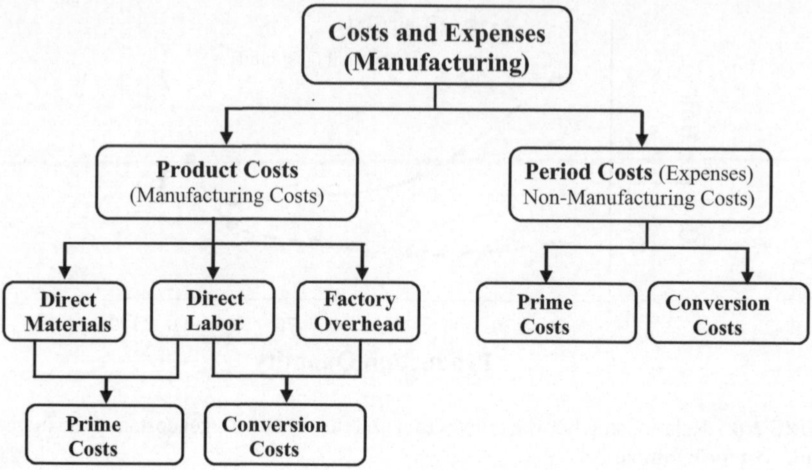

FIGURE 2.9 A Manufacturing enterprise's product costs and period costs (expenses).

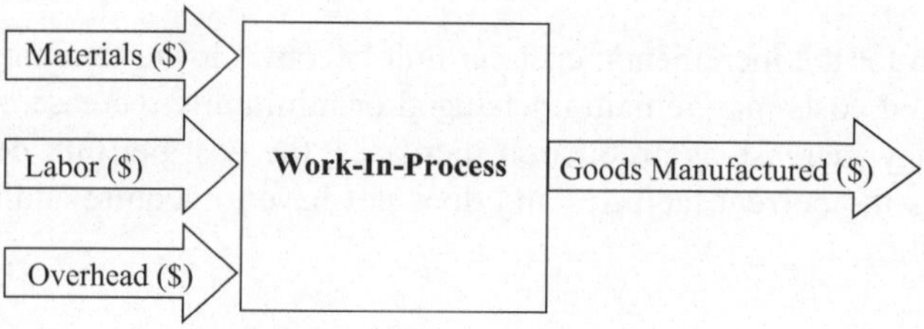

FIGURE 2.10 An illustration of cash flow (costs) for manufacturing goods.

Product Costs: Product costs (or inventoriable or manufacturing costs) are all the costs associated with the manufacturing of products. Normally, product costs include direct labor, direct materials (include freight-in), and factory overhead (Figure 2.10).

Period Costs: Period costs (also known as *commercial* costs) are the expenses that a firm incurs for its non-manufacturing general operations and sales activities. They include the administrative expenses, selling and distribution expenses, as well as depreciation expenses for non-manufacturing buildings and equipment. The period costs appear in a company's income statement for the fiscal period in which they incur.

2.4.5. Classification of Costs by Cash Flow

A firm may pay for costs incurred in a form of payment or accounting for the value of resources used. Therefore, one may classify the costs as cash costs, book costs, implicit (imputed) costs, and explicit costs.

Cash Costs: Cash costs (or out-of-pocket costs) are those that require the use of current resources, usually cash or an alternative form of payment (e.g., notes or loan payable, and accounts payable). For example, if a company purchases an asset in cash or on credit. Cash costs are the before-tax cash flows, such as cash outlays for a project's initial cost, annual operating and maintenance costs, and retirement/termination costs. They also include cash outlays due to financing (e.g., payments on principal and interest) and taxes.

Book Costs: Book costs are those costs that do not involve a cash exchange; instead, they only appear on the accounting books of the firm. For example, depreciation costs are not before-tax cash flows. Book costs do not represent before-tax cash flows. However, engineering economic analyses consider the book costs of depreciation expenses as they reduce the amount of taxes paid to governments.

Implicit Costs: An implicit (also known as *imputed* or *implied*) cost is an invisible cost that does not actually incur directly but is associated with internal transactions when a firm uses an internal resource (e.g., building, equipment, or owner's time) for a particular application, rather than using the resource to a different application. For example, when a firm can rent out a building or piece of equipment to an external party and earn income but uses it for an internal application, the estimated rent is an imputed cost. Also, when a firm transfers the work-in-process from one production/operating department to another, it may use a method of transfer costing, for budgetary reasons, and charge the assessed costs to the next department. Although there is no actual purchase of materials, the firm charges the receiving department with these imputed costs for the work it has received.

Explicit Costs: Explicit costs are those incurred due to a firm's activities and it pays for them in a form of payment. This makes the explicit costs equal to out-of-pocket costs.

2.4.6. Classification of Costs by Time

A firm may use the costs incurred in the past to estimate its planned production or, instead, it may use a predetermined costs rate, as described below.

Historical Costs: Historical costs are costs incurred in the past. They are sunk costs.

Predetermined Costs: Predetermined costs are the computed costs in advance on some basis of rate factors assigned to products.

2.4.7. Classification of Costs by Other Basis

Recurring Costs: Recurring costs (or *normal* costs) are those that occur on a regular basis as anticipated. For example, the costs of materials and labor used in manufacturing a product, utilities used by a production facility, property taxes, and insurance premiums are recurring costs.

Nonrecurring Costs: Nonrecurring costs (or *abnormal* costs) are those that do not occur on a regular basis (generally not anticipated). For example, the costs of replacing a production machine, the extra compensation (severance package paid for terminated employees), and a loss in the factory due to fire are nonrecurring costs.

Life-cycle Costs: Life-cycle costs are all costs that occur over the various phases of a product life cycle, from needs assessment through design, production, operation, and retirement of the product.

Target Costing: Target costing is the process of determining the desired cost for a product considering a given competitive price so that the product will generate the desired profit.

2.5. INVENTORY VALUATION

The price or production cost of a particular merchandise may vary during the year, and it is usually purchased at different points in time during the year and at different prices. In addition, when identical items are bought and sold, it is often impossible to tell which items have been sold and which are still in inventory. Therefore, it is necessary to make an assumption about the order in which items have been sold. Since the assumed order of sales may or may not approximate the actual order, the assumption is, in reality, about the flow of costs rather than the flow of physical inventory. Thus, the term *goods flow* refers to the actual physical movement of goods in the operations of the company, and the term *cost flow* refers to the association of costs with their assumed flow within the operations of the company. Therefore, the assumed cost flow may or may not correspond to the actual goods flow.

Inventory may be measured (appraised) using one of the four commonly practiced methods.

- **Specific Identification Method**: Under this method, the actual cost of a particular item is assigned to the item. However, the application of the specific identification method is limited to certain situations, particularly when the items have unique specifications or are very important and costly. For example, the inventory of jet engines should be measured by this method.
- **Average-Cost Method**: In this method, it is assumed that each item carries an equal cost, which is determined by dividing the total cost of the goods available for sale by the number of units to arrive at an average unit cost.
- **First-In, First-Out (FIFO) Method**: Under the FIFO method, it is assumed that the costs of the first items purchased are assigned to the first items sold and the costs of the last items purchased are assigned to the items remaining in inventory.
- **Last-In, First-Out (LIFO) Method**: Under the LIFO method, it is assumed that the costs of the last items purchased are assigned to the first items sold. As a consequence, the cost of the inventory on hand is composed of the costs of items from the oldest purchases.

Example 2.5 (Inventory valuation)

To illustrate the four methods, the following data for the month of June will be used:

Inventory Data, June 30

Date	No. of units	Unit cost	Total cost
June 1 (On hand)	100	$1.00	$100
June 6 (Purchase)	100	$1.10	110
June 13 (Purchase)	100	$1.20	120
June 20 (Purchase)	100	$1.30	130
June 25 (Purchase)	100	$1.40	140
Totals	500		$600
Sales	280		
June 30 (On hand)	220		

Using the Specific Identification Method. If the units in the ending inventory can be identified as coming from specific purchases, they may be priced at the specific prices of these purchases. For instance, assume that the June 30 inventory consisted of:

- 50 units of the June 1 inventory
- 100 units from the purchase of June 13, and
- 70 units from the purchase of June 25.

The cost to be assigned to the inventory under this method would be $268, determined as follows:

Inventory, June 30
(Specific Identification Method)

June 1	50 units	@ $1.00	$ 50
June 13	100 units	@ $1.20	120
June 25	70 units	@ $1.40	98
Total	220 units at a value of		$268

The cost of goods sold during June under the specific identification method is determined as follows:

Cost of Goods Available for Sale	$600
Less: Ending Inventory	268
Cost of Goods Sold	$332

Using the Average-Cost Method. Under this method, the average cost is computed by dividing the total cost of goods available for sale by the total units available for sale. That is,

$$\text{Average unit cost} = \$600 \div 500 \text{ units} = \$1.20 \text{ per unit}$$

$$\text{Ending Inventory} = 220 \text{ units} \times \$1.20 \text{ per unit} = \$264$$

The cost of goods sold during June under the average-cost method would be as follows:

Cost of Goods Available for Sale	$600
Less: Ending Inventory	264
Cost of Goods Sold	$336

Using the First-In, First-Out (FIFO) Method. When the FIFO method is used, the June 30 (ending) inventory would be $294. It is computed as follows:

Inventory, June 30
(FIFO Method)

From purchases of June 25	100 units	@ $1.40	$140
From purchases of June 20	100 units	@ $1.30	130
From purchases of June 13	20 units	@ $1.20	24
Total	220 units at a value of		$294

The cost of goods sold during June under the FIFO method would be $306, determined as follows:

Cost of Goods Available for Sale	$600
Less: Ending Inventory	264
Cost of Goods Sold	$306

Using the Last-In, First-Out (LIFO) Method. Under this method, the June 30 inventory would be $234, computed as follows:

Inventory, June 30
(LIFO Method)

From inventory at June 1	100 units	@ $1.00	$100
From purchases of June 6	100 units	@ $1.10	110
From purchases of June 13	20 units	@ $1.20	24
Total	220 units at a value of		$234

The cost of goods sold during June under the LIFO method would be $336, computed as follows:

Cost of Goods Available for Sale	$600
Less: Ending Inventory	234
Cost of Goods Sold	$366

Now, let us compare the effects of the four methods on the net income. Assume that sales during the month of June is $500.

	Specific Identification	Average Cost	FIFO	LIFO
Sales	$500	$500	$500	$500
Beginning Inventory	$100	$100	$100	$100
Purchases	500	500	500	500
Cost of Goods Available for Sale	$600	$600	$600	$600
Less: Ending Inventory	268	264	294	234
Cost of Goods Sold	$332	$336	$306	$366
Gross Profit on Sales	$168	$164	$194	$134

Keeping in mind that in this illustration June was a period of rising prices, it can be seen that LIFO, which charges the most recent (and in this case, the highest) prices to the cost of goods sold, resulted in the lowest net income. On the other hand, FIFO, which charges the earliest (and in this case, the lowest) prices to the cost of goods sold, produced the highest net income. The net income under the average-cost method is between those computed under LIFO and FIFO.

During a period of declining prices, the reverse effects would occur. The LIFO method would produce a higher net income than the FIFO method.

Each method has its own advantages and disadvantages, and none of them can be considered as best or perfect. The factors that should be considered in choosing an inventory method are the effects of each method on the balance sheet, income statement, income taxes, and management decisions.

The LIFO method is best suited for the income statement because it best matches revenues and the cost of goods sold. But it is not the best measure of the current balance sheet value of inventory, particularly when there has been a prolonged period of price rises or decreases. Many businesses use LIFO to reduce the amount of income taxes to be paid.

The FIFO method, on the other hand, is best suited to the balance sheet because the ending inventory is closest to current values and thus gives a more realistic view of the current financial assets of a business.

2.6. STATEMENT OF COST OF GOODS SOLD

To establish a foundation for the necessity and concepts of product costing, one must understand the flow of costs through a manufacturing organization. The "Cost-of-Goods-Sold Statement" reveals the flow of costs from raw material, which goes through the manufacturing process, and then transferred to finished goods inventory. In the period the finished goods are sold, their costs are transferred out of finished goods inventory to the cost of goods sold. However, the flow of costs for a merchandising company occurs somewhat differently. Costs and units move through material, work-in-process, and finished goods inventories as follows:

$$\textbf{Beginning Balance}\ (\textbf{BB}) + \textbf{Transfers In}\ (\textbf{TI}) - \textbf{Transfers Out}\ (\textbf{TO})$$
$$= \textbf{Ending Balance}\ (\textbf{EB})$$

A merchandising company buys and sells goods that are ready for resale when it receives them. The company does not need to do much to the goods to make them salable other than preparing a special package or displaying the items. Inventory cost generally includes the invoice price (less cash discount and purchases returns), freight and transportation (including insurance in transit), and applicable taxes and tariffs. Therefore, preparation of the cost of goods sold for a merchandising company is usually a simple procedure.

The cost-of-goods-sold statement is an important statement to the cost engineers. This statement is useful in determining the cost of manufacturing a particular product during a given accounting period, for which the statement is prepared. In this statement, all factory costs are entered. Figure 2.11 illustrates a general format of the cost-of-goods-sold statement.

The cost-of-goods-sold statement can be trimmed, by eliminating the "change in Finished Goods Inventory" and the "Cost of Goods Sold" (i.e., the bottom part), to get the "**Cost-of-Goods-Manufactured Statement**."

McAlter Corporation Statement of Cost of Goods Sold For Year Ended December 31, 20XX		
Materials:		
Beginning Inventory (1/1/20XX)	$ 30,000	
Purchases during the year	580,000	
Total Materials available for use	**$610,000**	
Ending Inventory (12/31/20XX)	20,000	
Cost of Materials used		$590,000
Direct Labor		390,000
Factory Overhead		95,000
Total Factory Cost during the year		**$1,075,000**
Change in Work-In-Process Inventory:		
Beginning Inventory (1/1/20XX)	$ 40,000	
Ending Inventory (12/31/20XX)	35,000	
Decrease in Work-In-Process Inventory		5,000
COST OF GOODS MANUFACTURED		**$1,080,000**
Change in Finished Goods Inventory:		
Beginning Inventory (1/1/20XX)	$ 33,500	
Ending Inventory (12/31/20XX)	98,500	
Increase in Finished Goods Inventory		(65,000)
COST OF GOODS SOLD		$1,015,000

FIGURE 2.11 A general format of the cost-of-goods-sold statement.

Example 2.6 (Cost of goods sold for merchandising firms)

The following is an example to illustrate the cost of goods sold computation in a hypothetical merchandising situation.

Merchandise Beginning Inventory	$2,000
Plus: Net Purchases (transferred in)	8,000
Cost of Goods Available for Sale	$10,000
Less: Merchandise Ending Inventory	2,700
Cost of Goods Sold (transferred out)	$7,300

Figure 2.12 visualizes the cost flow (or goods flow) through a merchandising company.

Accountants and cost analysts need information regarding the cost of raw materials placed into the production process, cost of goods manufactured, and cost of goods sold to prepare an income statement, appraise inventories, and prepare a balance sheet for a manufacturing company.

The flow of costs through a manufacturing operation is more complex than that in a merchandising company. In a manufacturing company, inventories are of three types: raw materials, work-in-process (partially completed products), and finished goods. As illustrated in Figure 2.13, a manufacturing firm buys raw materials and adds them to the beginning Raw Materials Inventory to compute raw materials available for use. The firm withdraws materials from this inventory and transfers them to the Work-In-Process Inventory, and it reports the cost of remaining unused materials on the company's balance sheet at the end of its fiscal period, as Raw Materials Ending Inventory. The firm adds the costs of direct labor and overheads to the beginning Work-In-Process (WIP) Inventory. It transfers the completed units and adds their costs to the Finished Goods Inventory and reports the costs of the partially completed units in WIP as Work In Process Inventory on its balance sheet. The firm deducts the cost of units sold (as an expense) from the revenues on its income statement and reports the costs of the unsold units on the balance sheet as the Finished Goods Ending Inventory (a current asset account).

The "**Cost-of-Goods-Manufactured Statement**" is similar to the Cost-of-Goods-Sold Statement, except it does not contain the part of the "Change in Finished Goods Inventory." For example, Figure 2.14 shows the cost of goods manufactured (transferred out) with the value of $1,080,000 for McAlter Corporation, as previously was shown in Figure 2.11.

2.7. JOURNAL ENTRIES FOR THE FLOW OF PRODUCTION COSTS

Consider the cost flow shown in Figure 2.15 based on which the corresponding journal entries are given in Figure 2.16 (wherein, the column for date is used for the referenced number, Ref).

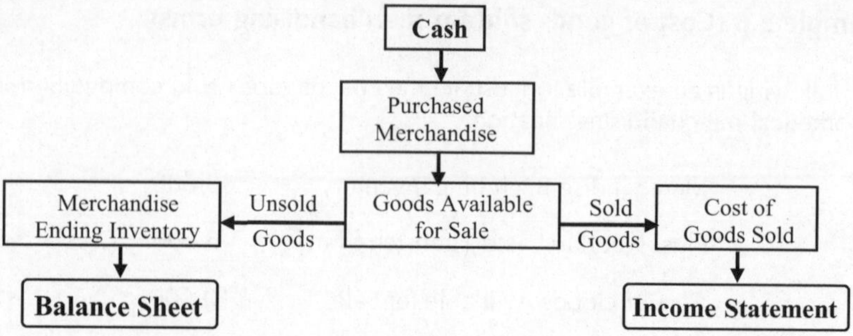

FIGURE 2.12 The flow of costs in a merchandising operation.

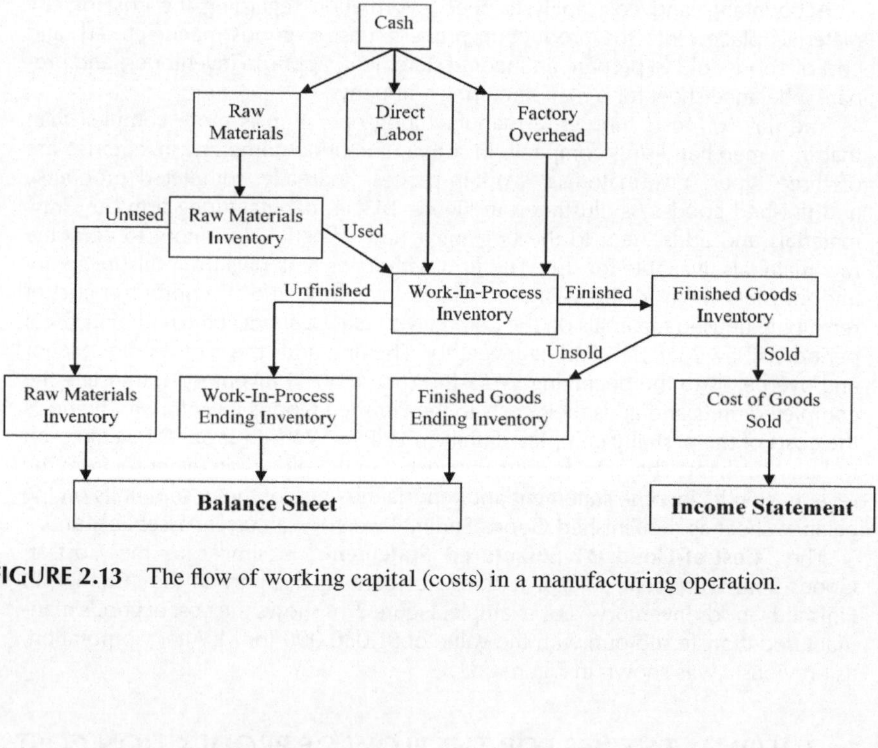

FIGURE 2.13 The flow of working capital (costs) in a manufacturing operation.

McAlter Corporation Statement of Cost of Goods Manufactured For Year Ended December 31, 20XX		
Materials:		
Beginning Inventory (1/1/20XX)	$30,000	
Purchases during the year	580,000	
Total Materials available for use	**$610,000**	
Ending Inventory (12/31/20XX)	20,000	
Cost of Materials used		$590,000
Direct Labor		390,000
Factory Overhead		95,000
Total Factory Cost during the year		**$1,075,000**
Change in Work-In-Process Inventory:		
Beginning Inventory (1/1/20XX)	$40,000	
Ending Inventory (12/31/20XX)	35,000	
Decrease in Work-In-Process Inventory		5,000
COST OF GOODS MANUFACTURED		**$1,080,000**

FIGURE 2.14 A general format for the cost-of-goods-manufactured statement.

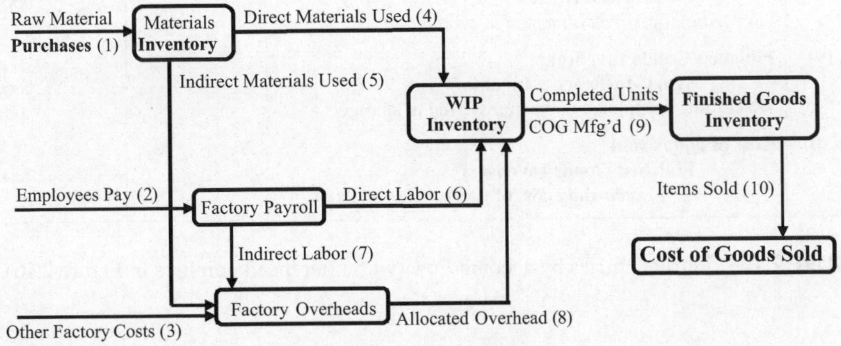

FIGURE 2.15 Production cost flowchart for journal entries (reference numbers in parentheses).

Ref	Account &Description	Debit	Credit
(1)	**Materials Inventory**	xxx	
	Cash or Accounts Payable		xxx
	Purchase of raw materials inventory		
(2)	**Factory Payroll**	xxx	
	Wages Payable		xxx
	Recording accrued wages, not yet paid		
(3)	**Factory Overheads**	xxx	
	Cash		xxx
	Recording factory hazard insurance		
	Factory Overheads	xxx	
	Cash		xxx
	Recording factory rent paid with cash		
	Factory Overheads	xxx	
	Accumulated Depreciation		xxx
	Recording depreciation on factory equipment		
	Factory Overheads	xxx	
	Supplies (Inventory)		xxx
	Recording supplies consumed by the factory		
(4)	**Work-In-Process Inventory**	xxx	
	Materials Inventory		xxx
	Recording direct materials used		
(5)	**Factory Overheads**	xxx	
	Materials Inventory		xxx
	Recording indirect materials used		
(6)	**Work-In-Process Inventory**	xxx	
	Factory Payroll		xxx
	Recording direct labor		
(7)	**Factory Overheads**	xxx	
	Factory Payroll		xxx
	Recording indirect labor		
(8)	**Work-In-Process Inventory**	xxx	
	Factory Overheads		xxx
	Recording overhead allocated to production		
(9)	**Finished Goods Inventory**	xxx	
	Work-In-Process Inventory		xxx
	Recording completed goods transferred to storage		
(10)	**Cost of goods sold**	xxx	
	Finished Goods Inventory		xxx
	Recording cost of goods sold		

FIGURE 2.16 Journal entries by account flow (with referenced numbers in Figure 2.16).

SUMMARY

This chapter presents the terminologies and fundamental concepts required for performing cost analysis. It provides cost classifications for various purposes and uses along with illustrative figures and computational examples where needed. It also presents the four common methods of inventory valuation and systematic procedures for computing cost of goods manufactured and cost of goods sold by formatted examples.

REFERENCE

Encyclopædia Britannica, Inc. (2020): *Diminishing returns*. Retrieved July 29, 2020 from: https://www.britannica.com/topic/diminishing-returns.

REVIEW QUESTIONS

R2.1. What is the formula for computing total variable cost?

R2.2. Describe the variable, fixed, and mixed costs.

R2.3. Describe the prime cost and conversion cost.

R2.4. What is the result of dividing the cost of goods manufactured by the number of units manufactured?

R2.5. The direct labor costs are added to the factory overhead costs to determine ___________.

R2.6. The total cost of goods manufactured is $900,000 for a company when produced 12,000 units of a product, of which it sold 10,000 units. Determine the product's manufacturing unit cost.

R2.7. A company is to decide whether or not to replace an old production machine with a new machine. Which of the following is not considered in the incremental analysis?

- Annual operating cost of the new machine.
- Annual operating cost of the old machine.
- Net cost of the new machine.
- Book value of the old machine.

R2.8. Explain why a company assigns product costs to its products and expenses to the period costs immediately in the period they incur.

R2.9. Arrange the following items in the order of the flow of cost of goods.

- WIP: work-in-process inventory.
- COGS: cost of goods sold.
- RMI: raw materials inventory.
- FGI: finished goods inventory.

R2.10. What is the difference between the cost of goods manufactured and the cost of goods sold?

PROBLEMS

P2.1. A factory incurs these costs for the month:

Direct materials	$25,000.
Direct labor	45,000.
Factory facility depreciation cost	30,000.
Factory utilities costs	2,500.
Staff's payroll salaries	22,000.

How much are the factory's prime costs and conversion cost?

P2.2. A small computer manufacturer produced 50,000 computers during the previous fiscal year. Their variable costs included $1,550,000 of direct materials (parts and supplies) and $1,200,000 of direct labor costs. The annual fixed costs totaled $2,000,000. Their projected sales volume for the next year is 60,000 computers. This sales increase will not change the rate of direct materials but will increase the fixed costs by $160,000. The company plans to give 2.5% raise to hourly workers who assemble the computers.

a) Compute the manufacturing unit cost during the previous year.
b) Estimate the manufacturing unit cost for the next year.

P2.3. Assume that a firm currently has a monthly total cost of $416,000 for manufacturing 16,000 units of its product. The management is considering expanding the production by 25%. The production expansion would increase the company's monthly costs to $480,000. The management wants to find the incremental costs involved.

P2.4. Polka Company can manufacture 10,000 computers per month with the following costs:

Direct material costs	$500,000
Direct labor cost	250,000
Variable overhead	100,000
Fixed overhead	150,000
Total Monthly Manufacturing Cost	**$1,000,000**

The company can purchase 10,000 computers from another company for $950,000. Of the total monthly fixed overhead costs, $50,000 are avoidable fixed costs if the computers are purchased externally. What effects would the purchase of the computers externally have on the A&F Company's costs?

P2.5. KML Company incurs the following costs for making 10,000 units of a component of its products:

Direct material costs	$ 9,600
Direct labor costs	10,200

Variable overhead costs	14,600
Fixed overhead costs	14,600
Total Cost	**$49,000**

The company can outsource manufacturing the component at $3.10 cost per unit. If the company takes this action, it could allocate its production capacity to increase the production of its other products that would generate an additional income of $4,600. Given that the fixed overhead costs are unavoidable, how much the company's net income would increase or decrease if it chooses the outsourcing option?

P2.6. The Maroon Company manufactures products A and B. Their **selling prices** of the products as they are partially and fully processed, and the required further processing costs to fully process them are as follows.

Product	Unfinished Price ($)	Finished Price ($)	Further Processing Costs ($)
A	70	125	45
B	85	220	160

Should any of the products be processed further? Why or why not?

P2.7. F&AT Company is selling a single product called TM4. The following data for this product during the year 20XX are available:

Date	Description	No. of Units	Unit Cost ($)
January 1	Beginning Inventory	17,000	6.00
February	Purchases	20,000	6.50
March	Purchases	40,000	6.20
May	Purchases	30,000	6.30
July	Purchases	50,000	6.40
September	Purchases	40,000	6.30
November	Purchases	15,000	6.60
Total Available for Sale		**212,000**	

Sales of product TM4 totaled 192,000 units at $10.00 per unit. Selling and administrative expenses totaled $592,000 for the year.

Required

1) Prepare a schedule to compute the cost of goods available for sale.
2) Prepare an income statement under each of the following assumptions:
   a) Costs are assigned to inventory on an average-cost basis.
   b) Costs are assigned to inventory on a FIFO basis.
   c) Costs are assigned to inventory on a LIFO basis.

P2.8. Rosan Company is merchandising a single product. The following information on beginning inventory, purchases, and sales for this product during the year 20XX is available

Date	Description	No. of Units	Unit Cost ($)
January 1	Beginning Inventory	15,000	9.90
March	Purchases	18,000	13.00
May	Purchases	20,000	12.50
July	Purchases	40,000	11.50
August	Purchases	40,000	13.00
October	Purchases	15,000	13.50
December	Purchases	12,000	15.00

Sales of the product totaled 130,000 units.

Required

Prepare a schedule to compute the cost of goods sold using the following methods:

a) Average-cost method.
b) First-In, First-Out (FIFO) method.
c) Last-In, First-Out (LIFO) method.

P2.9. The following data are available for Hee Haw Manufacturing Company. Prepare the Cost-of-Goods-Sold statement for this company.

	Inventories	
	01/01/20XX	12/31/20XX
Raw Materials	$ 50,000	$100,000
Work-In-Process	75,000	25,000
Finished Goods	100,000	150,000

The company has purchased $650,000 of raw materials and paid $400,000 as direct labor cost and $250,000 for factory indirect costs.

P2.10. Fragments of the operating data for a company for June and July are shown below. The company applies an overhead rate of 150% of direct labor cost. Fill in the blank with appropriate amounts.

	June	July
Raw Materials		
Beginning balance	(a) $_______	(b) $_______
Purchases	33,000	36,000
Requisitions	(35,000)	(c) (_______)

	June	July
Ending balance	$10,000	$14,000
Work-In-Process		
Beginning balance	$56,000	(d) $_____
Raw material requisitioned	(e) _____	32,000
Direct labor charges	20,000	18,000
Factory overhead applied	(f) _____	(g) _____
Total	(h) _____	(i) _____
Ending balance	(50,000)	(j) (_____)
Cost of Goods Manufactured	(k) $_____	$85,000
Finished Goods		
Beginning balance	$21,000	(l) $_____
Transfers from Work-In-Process	(m) _____	(n) _____
Total goods available for sale	(o) $_____	(p) $_____
Transfers to Cost of Goods Sold	(86,000)	(q) (_____)
Ending balance	(r) $_____	$30,000

P2.11. The following balances were extracted from the "Trial Balance" of BBY Manufacturing, Inc. at the end of the year 20XX. However, the Inventory accounts still reflect balances at the beginning of the year.

Raw Materials Inventory (1/1/20XX)	$505,000
Work-In-Process Inventory (1/1/20XX)	1,565,000
Finished Goods Inventory (1/1/20XX)	2,325,000
Purchases (Raw Materials)	4,752,000
Direct Labor Costs, Factory	2,210,000
Supplies Used, Factory	55,000
Depreciation, Machinery	60,000
Depreciation, Building, Factory	20,000
Depreciation, Building, Headquarters	31,500
Indirect Labor, Factory Services	842,000
Supervisory Salaries, Factory	280,000
Administrative Salaries, Non-factory	6,347,500
Sales Personnel Salaries and Wages	3,506,500
Property Taxes, Factory	60,000
Factory Miscellaneous Expenses	23,000
Advertising and Marketing Expenses	904,500

The year-end inventory balances are $457,000, $2,035,000, and $1,265,000 for Raw Materials, Work-In-Process, and Finished Goods Inventories, respectively.

Required

Prepare a statement (in good and organized form) to compute total manufacturing costs, cost of goods manufactured, and cost of goods sold for the year.

APPENDIX 2A

COST CLASSIFICATION

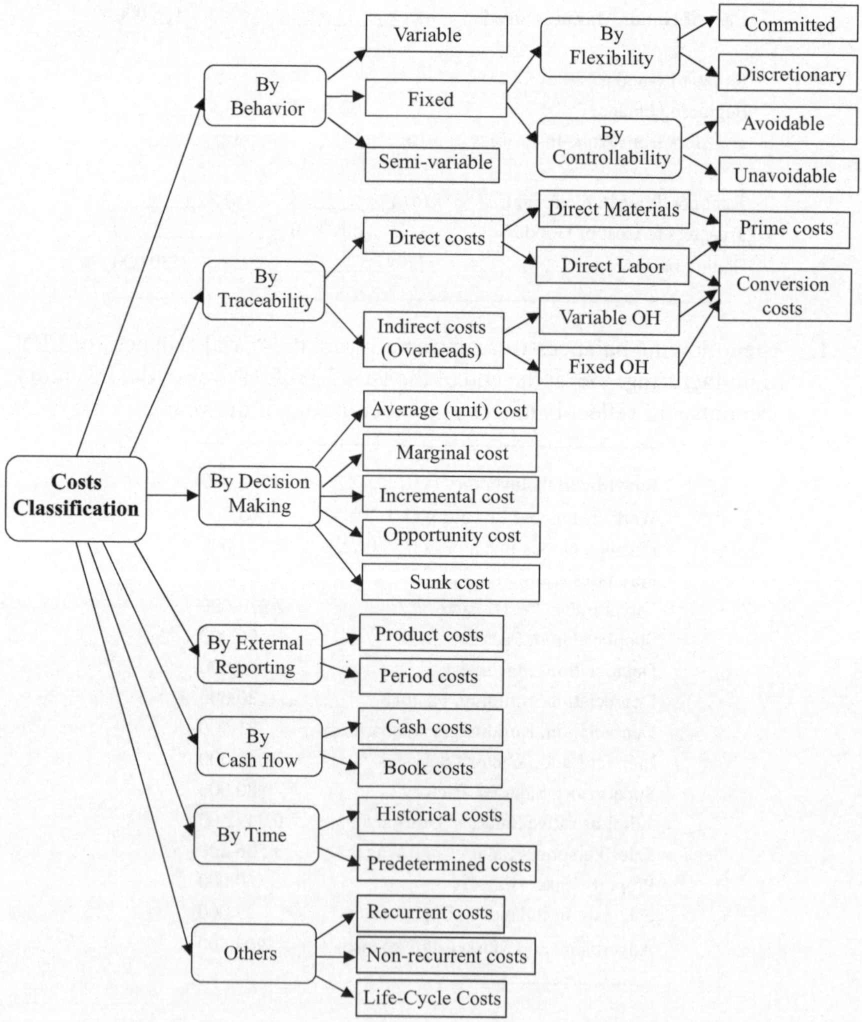

3 Product Costing

LEARNING OBJECTIVES

- Understand the overall cost flow from raw materials to finished goods sold throughout all stages within a firm.
- Understand the process for product costing.
- Be able to prepare the costs of goods manufactured and goods sold statement.
- Understand the general methods for a product, such as absorption, variable, job-order, and process costing.
- Understand the concept of equivalent units of production.
- Understand the overall cost flow from department to department, and how total inventory costs are compiled.
- Assign total costs to completed units and incomplete units still in process.

3.1. INTRODUCTION

Product costs consist of the costs of direct materials, direct labor, and overhead (or indirect costs). As presented in Chapter 2, the physical components of a product (i.e., direct materials) and the work performed (i.e., direct labor) to produce products are conveniently and economically traceable to the units of output, and assigning these direct costs is straightforward. On the contrary, overhead costs are not directly related to the production of products or providing services. As a result, these costs are not easily traceable to specific products unlike the direct costs. Hence, companies adopt a costing method for allocating the indirect costs to the cost objects (i.e., products, departments, jobs, etc.)

Costing refers to any process for assigning costs to an element of a business such as products, departments, employees, and customers. A product costing system is a costing process by which a firm estimates the costs of producing its products for managerial decision-making purposes, such as:

- Deciding what to sell.
- Product prices.
- Determining the cost of goods sold.
- Valuation of inventories.
- Finding potentials for cost reduction.

There are two main systems of product costing, **job-order** costing and **process** costing. The **job-order costing** (or **job costing**) system assigns costs to a specific cost object (i.e., a job, product, department, or customer). A firm assigns the costs of

DOI: 10.1201/9780429432163-3

materials to jobs consuming them according to the information on a materials requisition sheet (source document indicating the raw materials request for a specific production job) and has its employees track their time by jobs using a time ticket (as a source document indicating the number of hours an employee worked on specific jobs). It also allocates overheads to jobs using a predetermined overhead rate. A firm uses this costing method when individual products or batches of products are unique, and especially when jobs are being billed directly to specific customers or are expected to be audited by those customers.

Process costing is a method whereby, first, a firm accumulates its material, labor, and overhead costs for each department; and then, it allocates the total departmental cost to the individual product. Companies with continuous production processes use process costing, where they manufacture large quantities of the same product that passes through multiple cost centers (departments or stages of production), usually in long production runs.

Under both cost systems, companies usually apply manufacturing overheads to products or job orders using predetermined overhead rates, which are computed at the beginning of each fiscal period by dividing the estimated manufacturing overhead cost by an allocation base (also known as activity base or activity driver).

3.2. ABSORPTION COSTING AND VARIABLE COSTING

Absorption costing (also known as *full costing*) includes all manufacturing variable and fixed costs incurred in producing a product, while **variable costing** only includes the variable costs associated with the production of the product but not any fixed costs. Absorption costing is required under the Generally Accepted Accounting Principles (GAAP), established by the Financial Accounting Standards Board (FASB). The GAAP mandates publicly traded companies to use absorption costing for reporting to external stakeholders (e.g., investors, creditors, and tax agencies).

Variable costing will result in a lower breakeven point (or price) by showing a lower cost of goods sold (COGS). This, however, makes it more difficult to set proper pricing for a product. The variable costing reveals a higher gross profit margin as compared to the absorption costing, which reports much higher COGS. Nevertheless, the variable costing provides the management with a good foundation for the **cost-volume-profit** (CVP) analysis. A later chapter will present such an analysis in detail.

Since the most commonly used method of costing is absorption costing, this method is used in this chapter for product costing.

3.3. JOB-ORDER COSTING SYSTEM

Companies that produce special-order (or custom-made) products or services, or several different products, would usually use the job-order costing system. A job cost includes all the costs incurred for direct materials, direct labor, and overhead to complete a job. Direct materials and direct labor hours are conveniently traceable for a specific job. Overheads that are indirect costs are not easily traceable for a specific job.

When a firm starts a job and as it incurs costs (direct materials, direct labor, and applied overhead), its accounting department records them as a debit in a "Work-in-Process," WIP, or "Jobs-in-Process" account. When the job is finished, the accountants remove the job from the WIP by debiting the job costs to the "Finished Goods Inventory" account and crediting the WIP account.

The store personnel issue materials only upon receipt of a "Stores Requisition" form (Figure 3.1) signed by an individual who is authorized to request materials. The costs of direct materials are charged to each job according to the data on this form.

A time ticket or card (Figure 3.2), which shows how an employee's time was spent, is used to charge direct labor cost to specific jobs and indirect labor (e.g., maintenance) cost to WIP.

A firm may establish a predetermined overhead rate to budget overhead costs for jobs. The firm chooses a base for the predetermined overhead rate (e.g., direct labor costs, direct labor-hours, machine-hours, job size, and job volume). The formula for calculating a predetermined OH rate is:

$$\text{Predetermined OH Rate} = \frac{\text{Estimated Overhead}}{\text{Estimated Units in Allocation Base}\left(\text{Activity}\right)} \tag{3.1}$$

The firm allocates overhead costs to the jobs in WIP Inventory by taking the actual amount of the base multiplied by the predetermined overhead rate. At the end of the period or at the completion of a job, the firm compares the actual overhead costs to the applied overhead costs. If applied overhead is more than the actual overhead, the overhead is over-applied. If applied overhead is less than actual overhead, overhead is under-applied. Then, the firm will record adjusting entries on accounting books to correct the cost of goods sold. The formula for calculating the actual overhead rate is:

$$\text{Actual OH Rate} = \frac{\text{Actual Overhead}}{\text{Actual Units in the Allocation Base}\left(\text{Activity}\right)} \tag{3.2}$$

ABC Company

STORES REQUISITION

Requisition Number: **3D38** **Date**: May 10, 20XX

Issued by: John Roberton

Department: Machining

		For Store Personnel's Use	
Article	Quantity	Unit Cost	Amount
T17 Connector	6	$ 95.65	$573.90
B34 Housing with cover	3	145.75	437.25
			$1,011.15

Use: Direct Material **Job No. 3D38** Indirect Material ______

Authorized By: John Doe

Posted by: Brad Eliac

FIGURE 3.1 An illustration of a materials requisition form.

ABC Company Time Ticket					
Time Ticket #: 1036			**Date:** May 15, 20XX		
Employee: Martin Martini			**Station:** 6		
Started	**Ended**	**Hours Worked**	**Rate**	**Amount**	**Job Number**
8:00	12:00	4.0	$10.00	$40.00	**3D38**
12:30	3:00	2.5	10.00	25.00	**2H69**
3:00	4:00	1.0	10.00	10.00	**Maintenance**

FIGURE 3.2 An illustration of a time ticket form.

The following example illustrates how to compute a predetermined overhead rate.

Example 3.1

A manufacturing company, based on the costs incurred in the past three years, projects the overhead for the next year to be $279,000, as detailed below:

Indirect labor cost	$9,000
Indirect materials	28,000
Insurance	22,000
Factory utilities	65,000
Depreciation of equipment	75,000
Depreciation of building	63,000
Interest expense factory mortgage	17,000
Total	$279,000

The estimated direct labor cost in the upcoming year is $120,000. Using the direct labor cost as a basis (cost driver activity), calculate the predetermined overhead rate.

$$\text{Predetermined Overhead Rate} = \frac{\text{Estimated Overhead Cost}(\$279{,}000)}{\text{Estimated Level of Activity}(\$120{,}000)}$$

$$= \$2.325 \text{ per direct labor dollar}$$

$$(\text{or } 232.5\% \text{ of direct labor cost})$$

The criterion for choosing an allocation base (i.e., an activity base or cost driver) should be the "cause-and-effect relationship" in determining the predetermined overhead rate. The following are some possible drivers:

- Units produced.
- Direct labor-hours.

ABC Company Job Cost Sheet												
Job Number: Job Description:						Customer:						
		Direct Materials			Direct Labor			Applied Overhead				
Date	Description	Quantity	Unit Cost	Amount	Hours	Rate	Amount	Basis	Quantity	Rate	Amount	

Contract Date	
Date Started	
Date Completed	

Material Cost	
Labor Cost	
Applied OH	
Job Cost	

FIGURE 3.3 An illustration of a general job cost sheet form.

- Direct labor cost.
- Machine-hours.
- Direct material cost.
- Floor space occupied.
- Number of employees.

The job-order number and customer's information as well as all pertaining cost data (direct materials, direct labor, and applied overhead) are entered in a job cost sheet. Figure 3.3 is an example of a job cost sheet form, which neither is a standard form nor may be applicable to any particular company or industry sector. Companies design a job cost sheet that fits their particular needs.

3.4. PROCESS COSTING

Under the process costing systems, each production department prepares a process cost sheet for a given period, which usually shows some unfinished units either in its beginning WIP inventory, ending WIP inventory, or both. Therefore, it is necessary to determine a cost flow, that is, to determine the order in which costs are transferred into and out of the department.

Process costing begins with the concept of an equivalent number of complete units, ENCU (or simply EU), which is the amount of WIP inventory expressed in terms of the number of complete units. The cost per equivalent unit becomes the basis for the final cost allocated to WIP inventory and goods manufactured. A firm may use either the weighted-average-cost (also called average-cost) method or the first-in-first-out (FIFO) method for assessing the WIP inventory costs. A "production cost report" (or "cost of production report") provides an overall view of the entire process costing method.

3.4.1. Process Costing Using Weighted-Average Method

Under the weighted-average method of process costing, the analyst averages costs and uniformly applies them to both "completed and transferred-out units" and the "equivalent number of complete units" of unfinished units in the ending WIP inventory. Whereas, the FIFO process costing method assumes the units first entered a department are the ones first completed and transferred out. The **weighted-average process costing** method involves the following steps:

1) **Determine the physical (actual) units to account for**, that is, finding units in the WIP at the beginning of the period (from the end of the previous period), units started during the prior period, units completed and transferred out to the next department (or finished goods storage if the department is the final stage of the production), and unfinished units in WIP at the end of the period.
2) **Break up the costs into its components** (i.e., transferred-in [if any], direct materials, direct labor, and factory overhead costs).
3) **Determine the costs added during the current period for each cost component** (i.e., transferred-in, direct materials, direct labor, and factory overhead costs) from the preceding department and in the department.
4) **Compute the total costs to account for each cost component** (i.e., transferred-in, direct materials, direct labor, and factory overhead costs). That is, add the transferred-in cost of the units transferred in during the current period to the transferred-in cost of units in the beginning inventory of the WIP, and add the costs of direct materials, direct labor, and overhead incurred during the period to the corresponding cost in the beginning WIP balances for direct materials, direct labor, and factory overhead costs, respectively.
5) **Compute total equivalent units** (i.e., all the completed units plus the equivalent number of complete units in the ending WIP inventory). The equivalent units in the ending WIP are computed by the actual units multiplied by the percentages of their completion.
6) **Compute the cost per equivalent unit for each cost component** by dividing the total cost for each cost component (i.e., transferred-in, direct materials, direct labor, and OH) by the total equivalent units corresponding to the cost component.
7) **Allocating the cost between the units completed and transferred out and the units in the ending WIP inventory** by multiplying the sum of costs per equivalent unit of all cost components by the number of units

completed and transferred out, and multiplying the cost per equivalent unit for each cost components by the corresponding equivalent units in the WIP ending inventory.

Example 3.2

Compute the cost of goods manufactured and the cost components of the work-in-process ending inventory using the weighted-average method of process costing for the packaging department of KFA Company, which has the following data for July:

- A total of 20,000 units in WIP inventory at the beginning of the month with account balances of $95,000 for costs transferred in from the production department, and $5,660 direct materials, $4,320 direct labor costs, and $7,200 factory overheads. Note that these items form four cost components (i.e., transferred-in, direct materials, direct labor, and OH).
- A total of 180,000 units transferred in from the production department during the month at a total cost of $835,000.
- Costs added included direct material costs of $31,200, direct labor costs of $18,240, and factory overheads of $29,200.
- A total of 170,000 units completed and transferred (out) to the finished goods storage.
- The unfinished units in ending WIP inventory: 100% complete as for transferred-in cost, 80% complete as for materials, 60% complete as for direct labor cost, and 40% complete as for factory overheads.

Solution

The actual (physical) quantity:

WIP beginning inventory	20,000	units
Transferred In	180,000	
To be accounted for	**200,000**	**units**
Completed and transferred out	170,000	units
In WIP ending inventory	30,000	
Accounted for	**200,000**	**units**

Compute the equivalent units in the ending WIP inventory using the weighted-average method as follows:

Transferred in: 30,000 (100%) = 30,000 units equivalent
Materials: 30,000 (80%) = 24,000 units equivalent
Direct labor: 30,000 (40%) = 18,000 units equivalent
Factory overheads: 30,000 (80%) = 12,000 units equivalent

By adding the 170,000 units completed and transferred out to the individually computed equivalent units in the ending WIP inventory for each cost component, we obtain the total equivalent number of complete units for the corresponding cost element (row **e** below).

		Transferred-in	Direct Materials	Direct Labor	Factory Overheads
(*a*)	Units completed and transferred out	170,000	170,000	170,000	170,000
(*b*)	Units in ending WIP inventory	30,000	30,000	30,000	30,000
(*c*)	Percentage of completion in ending WIP	100	80	60	40
(*d*)	Equivalent units in ending WIP ($d = b \times c$)	30,000	24,000	18,000	12,000
(*e*)	Total equivalent units ($a + d$)	**200,000**	**194,000**	**188,000**	**182,000**

Compute the unit cost per equivalent using the weight-average method as follows: The costs to be accounted for is the sum of the cost of the beginning WIP inventory and the cost added during the period. We must compute this cost individually for the cost components as their equivalent units are not equal and we need them separately for cost control and managerial decision-making purposes.

Transferred-in cost: \$930,000/200,000 units = \$4.65 per unit
Direct material cost: \$36,860/194,000 units = \$0.19 per (equivalent) unit
Direct labor cost: \$22,560/188,000 units = \$0.12 per (equivalent) unit
Factory overhead cost: \$36,400/182,000 units = \$0.20 per (equivalent) unit
Overall manufacturing cost = 4.65 + 0.19 + 0.12 + 0.20 = \$5.16 per complete unit

		Transferred-in	Direct Materials	Labor Costs	Factory Overheads	Total
(*f*)	WIP beginning inventory	\$95,000	\$5,660	\$4,320	\$7,200	\$112,180
(*g*)	Added costs during the month	\$835,000	\$31,200	\$18,240	\$29,200	\$913,640
(*h*)	Costs to account for	\$930,000	\$36,860	\$22,560	\$36,400	\$1,025,820
(*i*)	Total equivalent units (*e*)	200,000	194,000	188,000	182,000	
(*j*)	**Cost per equivalent unit** (*h/e*)	**\$4.65**	**\$0.19**	**\$0.12**	**\$0.20**	**\$5.16**

Now, we can determine the cost of units completed and transferred to the finished goods storage as follows.

Cost of goods transferred out = \$5.16 × 170,000 = **\$877,200.**

We can also determine the cost of the 30,000 partially finished units in the ending WIP inventory as follows.

Transferred-in cost = \$4.65 × 30,000 = \$139,500
Material cost = \$0.19 × 24,000 = \$4,560
Direct labor cost = \$0.12 × 18,000 = \$2,160
Overhead cost = \$0.20 × 12,000 = \$2,400

We find the value of the ending work-in-process inventory to be:

$$\textbf{Work-in-process inventory} = \$139,500 = \$4,560 + \$2,160 + \$2,400$$
$$= \mathbf{\$148,620}$$

Since the cost of beginning WIP inventory plus the added cost during the period (cost to be accounted for) must be equal to the cost of units transferred out plus the cost of the ending WIP inventory (the costs accounted for), the cost of ending WIP can be determined by the following shortcut formula.

Cost of Ending WIP = Costs to Account For – Costs Transferred Out

In this example, we get the same amount for the ending WIP cost as follows:

$$\text{Cost of ending WIP} = \$1,025,820 - \$877,200 = \mathbf{\$148,620}.$$

At the end of each period, a production manager prepares a *"production cost report"* containing the information as presented in Figure 3.4. This report shows both the flow of units and the flow of costs through a processing center. It also shows how the costs are distributed between the units completed and transferred out and the units still in the processing center's ending inventory.

3.4.2. Process Costing Using FIFO Method

The FIFO method assumes any partially completed units from the previous period (beginning work-in-process inventory) are finished before the work on the units added during the current period starts. Table 3.1 clarifies the differences between the weighted-average method and the FIFO cost method.

Under the FIFO method, the process costing system calculates equivalent units for the following three items:

1) The units from beginning WIP: These units are completed first as the first-in, first-out requires, based on how much more materials, labor, and overhead will be needed to complete them. The percentage of the beginning WIP units to be completed during the current period equals 100% **minus** the percentage of completion in the previous period. So, the equivalent units from the beginning WIP equal **(physical units in beginning WIP) × (100% – the percentage of completion in the last period)**.

KFA Company
Production Cost Report
July 31, 20XX

Quantity Schedule:	
Beginning inventory units, July 1	20,000 units
Transferred-in units	180,000
Total units to be accounted for	**200,000** units
Transferred-out units	170,000
Ending inventory units, July 31	30,000
Total units accounted for	**200,000** units
Cost Schedule:	
Beginning inventory, Costs transferred in,	$ 95,000
Direct material costs	5,660
Direct labor costs	4,320
Factory overhead costs	7,200
Total beginning inventory costs	**$112,180**
Transferred-in cost during July	835,000
Direct material costs added	31,200
Direct labor costs added	18,240
Factory overhead costs added	29,200
Total costs to be accounted for	**$1,025,820**
Transferred out finished good	$877,200
Costs transferred-in, July 31	139,500
Direct material costs	4,560
Direct labor costs	2,160
Factory overhead costs	2,400
Total costs accounted for	**$1,025,820**

FIGURE 3.4 An illustration of the "production cost report" for Example 3.2.

2) The units started and completed in the current period: These units equal to total units completed in the current period **minus** the physical (actual) units in the beginning WIP.
3) The units remaining in ending WIP: As with the weighted-average method, we compute the equivalent units by taking the physical units' ending WIP multiplied by their percentage of competition.

The sum of these three equivalent units is the total equivalent number of units completed during the current period with the costs incurred during the period only (the beginning WIP costs are not included).

TABLE 3.1
Comparison of the Weighted-Average and FIFO Methods

	Weighted Average	FIFO
Units completed and transferred Out	Total units completed during the period	(All units in beginning WIP completed first) + (Units started and completed this period)
Equivalent units	(Units completed during the period) + (Equivalent units in ending WIP)	(Equivalent units in opening WIP completed) + (Units started and completed this period) + (Equivalent units in the ending WIP)
Cost per equivalent unit in ending WIP inventory	Beginning WIP costs + Costs added during the period	Costs added in current period only
Assign costs using	Equivalent units × (Cost per equivalent unit for units completed and units in ending WIP)	(Beginning WIP costs) + (Equivalent units) × (Cost per equivalent unit for units finished from beginning WIP, units started and completed, and units in ending WIP)

Therefore, we can formulate the computation for the equivalent units during a period under the FIFO as:

$$\begin{aligned}\mathbf{EU\,(or\;ENCU)} &= \left[\mathbf{U_B} \times (\mathbf{100\%} - \mathbf{P_L})\right] \\ &\quad + (\mathbf{U_T} - \mathbf{U_B}) \times (\mathbf{100\%}) + \left[\mathbf{U_E} \times \mathbf{P_E}\right]\end{aligned} \tag{3.3}$$

where

EU = ENCU during the period (separately computed for each cost component);

U_B = Actual (or physical) units in the beginning WIP inventory (carried over from the last period);

U_S = Units added (started) during the current period;

U_B + U_S = Total units to be accounted for;

U_T = Units completed and transferred out (to finished goods or the next stage) in the current period;

$U_T - U_B$ = Units started and completed in the current period;

U_E = Units in WIP inventory (partially completed) at the end of the current period.

U_T + U_E = Total units accounted for (= $U_B + U_S$ = Total units to be accounted for);

P_L = Percentage of the units in beginning WIP completed in the previous (last) period;

$(100\% - P_L)$ = Percentage of the units in beginning WIP completed in the current period;

$(U_S - U_E)$ = Units started and completed and transferred out during the current period;

P_E = Percentage of completion of the units in WIP at the end of the current period.

Example 3.3

Calculate equivalent units using the FIFO method for a production process with the following production data:

	Actual Units	% of Completion for Materials	% of Completion for Conversion
Units in beginning WIP (U_B)	20,000	80%	60%
Units added in the current period (U_S)	165,000	?	?
Units completed this period (U_T)	170,000	?	?
Units in ending WIP (U_E)	15,000	100%	60%
Cost of beginning WIP		$39,200	$13,200
Cost of units added this period		$425,880	$180,360
Total cost		$465,080	$193,560

Solution

	Direct Materials	Conversion Costs
Units in beginning WIP (U_B)	20,000	20,000
% of completion of beginning WIP in previous period (P_L)	80	60
% of beginning WIP completed this period [$P_B = 100\% - P_L$]	20	40
Equivalent units for beginning WIP [$U_B \times (100\% - P_L)$]	4,000	8,000
Units started and completed in current period ($U_S - U_B$)	150,000	150,000
Units closing WIP (*a*)	15,000	15,000
% of completion of closing WIP (*b*)	100	60
Equivalent units in closing WIP ($a \times b$)	**15,000**	**9,000**
Total equivalent units during the current period (EU)	**169,000**	**167,000**

Computing the unit cost per equivalent under the FIFO method

Under the weighted-average method, we use beginning work-in-process costs plus the costs added during the current period. Under the FIFO method, we only use the costs added during the current period. As we did under the weighted-average method, we must compute the unit costs individually for the cost components.

Direct material cost: $425,880/169,000 units = $2.52 per unit
Conversion cost: $180,360 /167,000 units = $1.08 per unit
Overall manufacturing cost = 2.52 + 1.08 = $3.60 per unit

Computation for assigning cost to units completed using the FIFO method

Direct material cost = $39,200 + $2.52(20,000 × 0.20 + 150,000) = $427,280
Conversion cost = $13,200 + 1.08(20,000 × 0.40 + 150,000) = $183,840
Cost of goods manufactured = $427,280 + $183,840 = **$611,120**

Computation for assigning cost to ending work-in-process inventory

Direct material cost = $2.52(15,000 × 1.00) = $37,800
Conversion cost = $1.08(15,000 × 0.60) = $9,720
Total cost assigned to ending WIP inventory = $37,800 + $9,720 = $47,520

We can use the following shortcut method to calculate the total cost of goods manufactured as follows.

$$\begin{aligned}\text{Total cost to be accounted for} &= \text{Cost of beginning WIP inventory} \\ &\quad + \text{cost added during the period} \\ &= (\$39,200 + \$13,200) + (\$425,880 + \$180,360) \\ &= \$658,640\end{aligned}$$

$$\begin{aligned}\text{Cost of Goods Manufactured} &= \text{Total cost to be accounted for} - \text{Cost Ending WIP} \\ &= \$658,640 - \$47,520 \\ &= \mathbf{\$611,1120}\end{aligned}$$

Example 3.4

A manufacturing company, based on the costs incurred in the past three years, projects the overhead for the next year to be $279,000 as detailed below.

Indirect labor cost	$9,000
Indirect materials	28,000
Insurance	22,000
Factory utilities	65,000
Depreciation of equipment	75,000
Depreciation of building	63,000
Interest expense factory mortgage	17,000
Total	$279,000

The estimated direct labor cost in the upcoming year is $120,000. Using the direct labor cost as a basis (cost driver activity), calculate the predetermined overhead rate.

$$\text{Predetermined overhead rate} = \frac{\text{Estimated overhead cost } (\$279{,}000)}{\text{Estimated level of activity } (\$120{,}000)}$$

$$= \mathbf{\$2.325} \text{ per direct labor dollar}$$

$$(\text{or } \mathbf{232.5\%} \text{ of direct labor cost})$$

3.5. NORMAL COSTING AND STANDARD COSTING

Normal costing is a process of computing the manufacturing costs of products, which consists of:

- The **actual** cost of direct materials.
- The **actual** cost of direct labor.
- The **applied** manufacturing overhead is based on a predetermined manufacturing overhead rate.

The normal product costs are used for determining the cost of goods manufactured, the cost of goods sold, and the cost of the inventories.

At the end of each fiscal period, the overhead variance, which is the amount of over-applied or under-applied overhead, is determined. If the amount of the applied overhead exceeds the actual overhead cost incurred, the variance is an over-applied overhead; if the amount of the applied overhead is less than the actual overhead cost incurred, the variance is an under-applied overhead.

If the amount of the variance is insignificant, it will usually be written off (entirely allocated) to the COGS account. That is, the COGS account will be debited for the under-applied overhead (added to COGS); it will be credited for the over-applied overhead (deducted from COGS).

If the variance is significant, it should be prorated to the cost of goods sold, the work-in-process inventory, and the finished goods inventory relative to the amounts of applied overhead to them.

Standard costing is a system used by some manufacturing firms to identify the differences (variances) between:

- The actual costs of the products manufactured, and
- The standard (budgeted or planned) costs that should have occurred for the products produced.

Standard costs are planned (or budgeted) costs that **should have occurred** for producing manufactured products, which consist of the following:

- Predetermined material costs.
- Predetermined direct labor costs.
- Predetermined manufacturing overhead costs.

These standard costs are used to determine the cost of goods sold and various inventories. Under the normal costing, there may be only overhead variance. Under the standard costing, there can be variances for all three cost components (direct materials, direct labor, and overhead). The variances are adjusted as done for the overhead variance under the normal costing. That is, if the actual costs vary insignificantly from the standard costs, the variances will be allocated (written off) to the cost of goods sold. Otherwise, they should be prorated to the cost of goods sold and various inventories relative to the amounts of the standard costs.

3.6. JOURNAL ENTRIES FOR RECORDING JOB-ORDER COSTS

This section presents typical journal entries for overhead costs. The recording can apply to both job-order and process costing systems. However, we focus on the job-order costing system as it requires more details.

The following journal entries record the direct material (DM) and direct labor (DL) costs as they are incurred:

	Debit	Credit
Work-In-Process-DM: Job J_1	6,000	
Work-In-Process-DM: Job J_2	4,000	
Materials Inventory		10,000
Work-In-Process-DL: Job J_1	3,000	
Work-In-Process-DL: Job J_2	2,000	
Wages Payable		5,000

The following journal entry is made when the wages (payable) are paid:

	Debit	Credit
Wages Payable	5,000	
Cash		5,000

The Factory Overhead account is debited as the actual overhead incurred, whereas the Work-In-Process (jobs or processes) account is debited and the Factory Overhead account is credited for the applied (allocated) overheads. Table 3.2 presents some examples of affected accounts for the actual overheads and Table 3.3 shows examples of affected accounts for the applied overheads.

The following is an example of a journal entry for recording the factory overhead costs incurred:

Date	Description	Debit	Credit
20XX-05-20	Factory Overhead	95,000	
	Supplies		12,000
	Salaries Payable		15,600
	Accumulated Depreciation		11,400
	Taxes Payable		5,000
	Utilities Payable		5,000

The applied (or budgeted) overhead for the period equals the actual hours worked times the predetermined overhead rate. Predetermined overhead rate is that a company estimates, at the period, how much overhead will cost per unit of an activity (e.g., labor hour). For example, a company estimates it will incur $10 per labor hour of overhead. The employees work 1,000 hours during the period. Therefore, applied (budgeted) overhead for the period equals $10,000 = $10 × 1,000 hours.

TABLE 3.2
Examples of the Affected Accounts for the Actual Overheads Incurred

Actual Overhead Item (example)	Debit	Credit
Indirect materials	Factory Overhead	Raw Materials or Supplies
Indirect labor (e.g., supervisory and HR)	Factory Overhead	Salaries Payable
Machineries depreciation cost	Factory Overhead	Accumulated Depreciation
Factory property taxes	Factory Overhead	Taxes Payable
Utilities	Factory Overhead	Accrued Expenses, or Utilities Payable
Equipment leases	Factory Overhead	Prepaid Lease

TABLE 3.3
An Illustration of How the Budgeted (Estimated) Overheads Are Allocated

Applied Overhead Item (example)	Debit	Credit
Allocated overheads to Job J_1	Work-In-Process	Factory Overhead
Allocated overheads to Job J_2	Work-In-Process	Factory Overhead
⋮	⋮	⋮
Allocated overheads to Job J_n	Factory Overhead	Factory Overhead

Under-applied overhead occurs when a company has the actual overhead costs greater than its applied overhead costs. Over-applied overhead, on the other hand, occurs when a company has the actual overhead costs less than its applied overhead costs.

To determine whether the overhead is under- or over-applied, an overhead analysis is necessary. The actual overhead costs are found through company recordings as they incur. However, if the actual overhead cost rates are known, then multiply those costs per unit of the activity (e.g., labor hour) by the number of activity units that occurred (e.g., labor hours worked). For example, if the actual overhead rate for the company is \$11 per hour, then the actual overhead is \$11,000 = \$11 × 1,000 hours. Subtracting the applied overhead costs from the actual overhead costs shows how the overhead has been applied. In our example, \$11,000 minus \$10,000 equals \$1,000 of under-applied overhead. If the applied overhead costs were greater than the actual overhead costs, then the overhead is over-applied.

The following is the journal entry to record the incurred indirect labor (e.g., managerial and supervisory salaries):

Factory Overhead	8,000	
Salaries Payable		8,000

The following is the journal entry to record the indirect materials raw materials are used:

Factory Overhead	1,000	
Materials Inventory		1,000

The following journal entry is to record to allocated (applied) manufacturing overhead cost to jobs (or departments):

Work-In-Process-OH – Job J_1	5,000	
Work-In-Process-OH – Job J_2	3,500	
Factory Overhead		8,500

The following journal entry records a completed job and transferred from the production department to the finished goods warehouse:

Finished Goods	23,500	
Work-In-Process – Job J_1		14,000
Work-In-Process – Job J_2		9,500

Note that the total cost of a job includes direct material and direct labor costs plus the allocated factory overheads.

As illustrated in the above journal entry examples and assuming they are the only required journal recordings, the Factory Overhead account in the general ledger would show the information as shown in Figure 3.5.

For our example, assume both jobs have been completed and shipped to the customers, where the sales amounts charged to customers were $18,000 for job J_1 and $13,000 for job J_2, totaling $31,000. The following journal entries record the **sales** and **cost** of the completed jobs and shipped to the customers:

Accounts Receivable	31,000	
Sales		31,000
Cost of Goods Sold	23,500	
Finished Goods		23,500

The Factory Overhead account is not a regular account like an asset, liability, expense, or any other typical accounts. It is a "**clearing**" account. All factory overhead costs enter this account and are then transferred out to other accounts. In this case, actual overheads go in (debited), and applied overheads go out (credited). The balance is either negative (over-applied overheads) or positive (under-applied overheads).

In our example, we realize that we incurred $9,000 factory overheads, but we have applied $8,500 to the jobs ($5,000 and $3,500 to jobs J_1 and J_2, respectively). Therefore, we have $500 under-applied factory overhead costs to the jobs. The following journal entry corrects an **under-applied** manufacturing overhead balance by adjusting the Cost of Goods Sold account:

Account Name: **Factory Overheads**

Date	Description	Ref.	Debit	Credit	Balance
mm/dd	Indirect labor		8,000		8,000
mm/dd	Indirect materials		1,000		9,000
mm/dd	OH Allocated to Job J_1			5,000	4,000
mm/dd	OH Allocated to Job J_2			3,500	500
⋮	⋮		⋮	⋮	⋮

FIGURE 3.5 An illustration of posted transactions in a job-order costing system.

Cost of Goods Sold	500	
Factory Overheads		500

Transfer the under-applied overhead to COGS

If the applied overheads exceed the actual overheads incurred, an **over-applied** overhead occurs. Then, for the over-applied amount, the Factory Overheads account is debited (to close it) and the Cost of Goods Sold account is credited (to reduce it to the actual amount). For example, if the total actual factory overhead is $9,000 and the applied overhead is $10,200, the following journal entry will be made:

Factory Overheads	1,200	
Cost of Goods Sold		1,200

To reduce COGS for the over-applied overhead

SUMMARY

In sequential production departments, units are typically transferred from one stage to the next stage in the process. When the units are transferred, the accumulated cost per unit is transferred along with them. Since the unit being produced includes work from all of the prior departments, the transferred-in cost is the cost of the work performed in all previous departments.

In any stage, the costs transferred in from the previous stage are added to the costs incurred during the current period, then the total of each cost component (i.e., the direct material, direct labor, and overhead costs) is divided by the equivalent units (units transferred out plus equivalent units in WIP) to determine the unit cost. The total unit cost in this stage is multiplied by the units completed during the current period and transferred out to the next stage to determine the total cost of the completed units and passed on to the next stage of the process. The costs assigned to the WIP ending inventory by taking the unit cost of each cost component (i.e., material, labor, and overhead cost) multiplied by the corresponding equivalent units in the WIP at the end of the period.

REVIEW QUESTIONS

R3.1. For what inventory accounts is the following equation true?

Beginning Balance + Transfers In – Transfers Out = Ending Balance

A. Raw Materials.
B. Work-in-Process.
C. Finished Goods.
D. All of the above.
E. B and C only.

R3.2. The predetermined rate is used to apply overhead costs to:

A. Raw materials inventory.
B. WIP inventory.
C. Finished goods inventory.

D. Work-in-process inventory and units transferred out.
E. Sales units.

R3.3. The costs to be accounted for consist of:
A. Costs added during the period.
B. Costs of the units in ending inventory.
C. Costs started and transferred out during the period.
D. Costs in the beginning inventory and added during the period.

R3.4. The detailed materials, labor, and overhead costs are needed when:
A. Determining the units to which costs are assigned.
B. Determining the equivalent units of production.
C. Determining the cost per equivalent unit.
D. Allocating the costs to the units transferred out and the units partially completed.

R3.5. How do job-order costing and process costing differ with respect to recording direct materials and direct labor?

R3.6. How is the predetermined rate calculated for applying the overhead costs to work-in-process?

R3.7. The companies that produce many different products or services usually use:
A. Process costing.
B. Job-order costing.
C. Both process costing and job-order costing.
D. None of the above.

R3.8. Which of the following costs is recorded on the job cost sheet?
A. Direct material cost.
B. Direct labor cost.
C. Manufacturing overhead cost.
D. All of the above.

R3.9. Which of the following journal entries is correct for the issuance of direct materials to production?
A. Materials debited and work-in-process credited.
B. Work-in-process debited and materials credited.
C. Work-in-process debited and accounts payable credited.
D. Materials debited and accounts payable credited.

R3.10. Production reports for the second quarter show the following data:

Month	Direct Labor Hours	Machine-Hours	Applied Overhead ($)
May	12,000	10,000	50,000
June	11,000	12,000	60,000
July	9,500	11,000	55,000

Which variable would be the most likely basis for allocating overhead?

PROBLEMS

P3.1. A production department within a company received materials of $12,000 and conversion costs of $13,000 from the prior department. It added materials of $32,800 and conversion costs of $50,000. The equivalent units are 20,000 for material and 18,000 for conversion. Determine the unit cost for materials and conversion, and the overall manufacturing unit cost.

P3.2. The finishing department received 5,350 units from the molding department and started the month with 650 units in WIP inventory. It transferred out 5,330 units to the packaging department during the month. How many units remain in process at the end of the month?

P3.3. A production department started a month with no WIP beginning inventory. It also started 60,000 units in production during the month, and completed and transferred out 56,800 units. Its remaining units in process at the end of the month were 100% complete for materials and 60% complete for conversion. Using the average cost method, compute the equivalent units of production.

P3.4. A production department started a month with 2,200 units WIP beginning inventory (100% complete for materials; 70% complete for conversion). It also started 60,000 units in production during the month, and completed and transferred out 56,800 units. The remaining units were still in process at the end of the month, which were 100% complete for materials and 50% complete for conversion. Using the average cost method, compute the equivalent units of production.

P3.5. In a production department, materials are added at the beginning of the process. There were 1,200 units in beginning WIP inventory, 11,800 units were started during the period, and 11,500 units were completed and transferred to finished goods inventory. The WIP ending inventory in the department was 60% complete as to conversion costs. Under the average cost method, what are the equivalent units of production for materials and conversion?

P3.6. The following data relate to work-in-process for a production department during a given month:

Work-in-process – beginning inventory:	
Units in process	1,200
Material cost (100% complete)	$6,600
Conversion cost (50% complete)	$2,400
Units added during the period	10,000
Costs incurred in this period:	
Direct materials used	$53,880
Direct labor cost	$18,800
Overhead	$23,000
Units of product transferred out	9,200

The unfinished units in the WIP ending inventory are 80% complete as for material costs and 60% complete as for conversion costs.

Required: Using these data, compute:

a) The unit cost per equivalent unit for materials and conversion (use the average cost method).
b) The cost of the product transferred out.
c) The cost of ending WIP inventory.

P3.7. The Abdo company has the following estimated costs for the next year:

Direct materials	$21,000
Indirect materials	$7,000
Direct labor	$46,000
Salary of production supervisor	$33,000
Rent on factory equipment	$14,000
Sales commission	$36,000
Advertising expenses	$9,000

It is estimated that 35,000 machine-hours and 10,000 direct labor hours will be worked during the next year.

Required: What will be the predetermined overhead rate if the company applies manufacturing overhead cost to jobs on the basis of direct labor hours.

P3.8. Balki Co. uses a predetermined overhead rate based on direct labor hours to apply manufacturing overhead to jobs. For April, the company's estimated manufacturing overhead cost was $200,000 based on an estimated activity level of 50,000 direct labor hours. The actual overhead amounted to $208,000 with actual direct labor hours totaling 55,000 for the month. How much was the over-applied or under-applied overhead?

P3.9. The over-applied balance of the Factory Overhead ledger account is $25,000. The ending balances of Work-in-Process inventory, Finished Goods inventory, and Cost of Goods Sold accounts are $5,000, $8,000, and $37,000, respectively. On the basis of ending balances, how much of the over-applied balance should be proportionated to each of these accounts?

P3.10. Production reports for the second quarter show the following data:

Month	Direct Labor Hours	Machine-Hours	Applied Overhead ($)
May	12,000	10,000	50,000
June	11,000	12,000	60,000
July	9,500	11,000	55,000

Which variable would be the most likely basis for allocating overhead?

APPENDIX 3A

GLOSSARY

Actual Overhead Rate: Total actual manufacturing overhead divided by total actual manufacturing activity.

Cost: A financial measure of the resources used or given up to achieve a purpose.

Cost Driver: Activity or transaction that causes costs to incur. For example, machine-hours can be a cost driver for costs of electricity and fuel to run machines.

Cost of Goods Manufactured: Consists of the total costs of all goods completed during the period; it includes all manufacturing costs incurred during a period plus the beginning work-in-process inventory costs minus ending work-in-process inventory costs.

Cost of Goods Sold: Cost of goods manufactured plus the beginning finished goods inventory minus the ending finished goods inventory.

Direct Labor: Labor costs of all employees directly engaged in converting materials into finished goods. Direct labor costs are clearly and conveniently traceable to specific products.

Direct Materials: Materials whose only use is to make specific products and their costs are clearly and conveniently traceable to those products.

Finished Goods: Manufactured products that are ready for sale.

Finished Goods Inventory: The title of an account for maintaining a record for finished products.

Indirect Costs: All costs of producing goods or providing services except for the costs of direct materials and direct labor.

Indirect Labor: Employees who work to support the production process. However, they do not take an active part in converting the materials into finished goods. Maintenance staff, security, custodians, human resources staff, and accountants are some examples of indirect labor.

Indirect Materials: Materials that a firm uses in producing the products but are not easily traceable to specific products or jobs. Examples include glue, oil, fuel, tapes, and cleaning supplies.

Job-Order Cost System (Job Costing System): A method for accumulating all costs incurred for producing a specific product according to the individual jobs, such as a building, a consulting job, or a batch of 100 desks.

Manufacturing Overheads: See Indirect Costs.

Materials: All items used in manufacturing processes.

Over-Applied (Over-Absorbed) Overhead: The amount by which the overhead applied to production in a period exceeds the actual overhead costs incurred in that same period.

Overheads: See Indirect Costs.

Period Costs: Costs not related to the production of products but incurred for the general operations of a firm. They include administrative, selling, and interest expenses.

Predetermined Overhead Rate: Determined by dividing estimated total overhead costs for a period by the estimated level of activity, such as estimated total machine-hours or direct labor-hours for the period.

Product Costs: Costs a company assigns to units produced. In manufacturing companies, these costs are direct materials, direct labor, and manufacturing overhead. In service companies that have no materials, these costs are direct labor and overhead.

Statement of Cost of Goods Manufactured: An accounting report showing the cost to manufacture and the cost of goods manufactured.

Total Manufacturing Costs: Includes the direct materials, direct labor, and manufacturing overhead costs incurred during a period.

Under-Applied (Under-Absorbed) Overhead: The amount by which actual overhead costs incurred in a period exceed the overhead applied to production in that period.

Work-in-Process: Partially completed products at the end of a period; a firm records and reports such incomplete products as Work-in-Process Inventory.

4 Manufacturing Cost Allocation

LEARNING OBJECTIVES

- Describe the cost allocation process.
- Compute predetermined overhead rates.
- Calculate departmental overhead rates.
- Describe the difference between support departments and operating departments.
- Allocate support department costs to operating departments using the direct, step-down (sequential), and reciprocal methods.
- Distinguish between plantwide overhead rates, departmental overhead rates, and activity-based overhead rates.

4.1. INTRODUCTION

Costs are the necessary expenditures that a business must spend to produce products or provide services. Providing goods or services or undertaking projects have economic consequences (costs and benefits). **Costing** is the process of determining costs of all resources (capital, human, and equipment) consumed in producing and delivering the products and services or completing the projects. Whereas, **valuation** is the process of determining the worth of the products, services, or projects.

A proper cost allocation is a crucial process for a firm to accurately compute the costs for all product lines or departments and, accordingly, to determine the per-unit costs and profits. With such information, a firm would have insights for improving the profitability of certain products, or developing strategies for cost reductions. However, misallocation of costs may cause making incorrect decisions. Cost analysts should make sure that they correctly allocate costs to the affected cost objects and that they choose appropriate cost allocation bases.

The direct (materials and labor) costs may conveniently be traced and assigned to specific products or operating departments. However, the overheads (costs of service departments) may not be traceable to specific products or operating departments. Therefore, a firm must use an appropriate method to allocate the overheads to its cost objects (e.g., product lines, departments, or customers).

A firm needs a system to track the costs of its operations. Two of the most commonly used systems are **traditional costing** and **activity-based costing**. Of these, one of these is easy to use and inexpensive, while the other requires more time and effort (costlier) to implement but provides superior accuracy.

DOI: 10.1201/9780429432163-4

A firm should group the individual costs into cost pools, from which it allocates the costs. Maintenance costs, warehouse costs, human resources costs, safety and security costs, and custodial costs are some typical examples of cost pools. A firm usually uses a single cost allocation basis, such as labor hours, machine hours, the number of employees, or occupied floor space, to allocate costs from a cost pool to the affected cost objects.

4.2. TYPES OF DEPARTMENTS

In a manufacturing firm, there are two types of departments, ***operating*** departments and ***service*** departments. The operating departments carry out the main objectives of the firm, which are producing products or providing services to clients. These departments are directly engaged in the transformation of raw materials or assembling the parts into finished goods in a manufacturing organization, or the provision of services to clients in a service organization. The following are examples of **operating/production departments**:

- Assembling
- Casting and molding
- Cutting
- Electronics
- Fabricating
- Finishing
- Joining
- Knitting
- Machining
- Matching
- Milling
- Mixing
- Packaging
- Painting
- Planing (woodworking)
- Refining
- Shearing and forming
- Upholstery

The service departments are not directly engaged in manufacturing goods or providing services. However, they provide services to operating departments. The following are examples of **service departments**:

- Accounting services
- Cafeteria
- Custodial services
- Engineering and Design
- Human resource
- Information and Technology
- Inspection department
- Internal auditing department
- Legal support
- Maintenance and Repairs
- Medical services
- Production scheduling
- Purchasing
- Quality assurance
- Receiving
- Research and Development
- Parking and landscaping
- Safety and Security
- Shipping
- Supplier relations
- Warehouse (storage)

4.3. TRADITIONAL OVERHEAD COST ALLOCATION

Recall that the total cost of a product is the sum of its direct materials, direct labor, and share of manufacturing overhead costs. A firm can conveniently assign the ***direct material costs*** to cost objects (i.e., products, jobs, or departments) using its preferred inventory valuation method. It can also conveniently assign the ***direct labor costs*** to the cost objects by determining the labor hours used by each cost object and multiplying them by the labor cost rates.

Costs allocation serves three main purposes: (1) make decisions, (2) reduce non-value-added costs, and (3) determine pricing. Managing cost control and evaluation of various departments, the management needs a systematic process for allocating the overhead (indirect) costs to multiple products or multiple departments. A typical cost allocation process in a multi-department company includes the following steps:

1. Allocating the overhead costs to all cost objects/centers (i.e., production/operating and service departments);
2. Reallocating the costs of service departments to the production/operating departments;
3. Finally, assigning the costs of each production department (including the allocated costs of service departments) to product lines.

Another challenge in costing is the adequacy of allocation of the overhead costs to specific cost objects (departments, products, or jobs). Choosing a reasonable method is the key to achieving this objective.

Cost analysts may use one of the two general common methods: "***predetermined overhead rate***" or "***overhead accumulation***."

4.3.1. Predetermined-Overhead-Rate Method

One of the most common approaches for assigning overhead costs is developing and using a predetermined overhead rate. Such a rate is an overhead cost factor for assigning factory overhead costs to specific units or jobs. A firm determines the rate based on projected overhead costs and production volumes for the period. This method is suitable for job-order costing, and for cases in which price determination is necessary at the time the job orders are taken. The rate is computed in three steps:

1) Using cost behavior analysis, estimate all overhead costs for the coming fiscal period.
2) Select an appropriate measure as the allocation basis, which can closely relate overhead costs to the products produced. The most common measures for cost allocation are direct labor hours, direct labor dollars, machine hours, units of output, number of requisitions, and floor space. The selected allocation measure becomes the denominator of the fraction used for computing the predetermined overhead rate.
3) Divide the total estimated overhead costs for the period by the total basis (hours, dollars, units, square feet, etc.) expected for the period. The result is a predetermined overhead rate, as expressed by the following formula.

$$\frac{\text{Predetermined overhead rate}}{\left(\text{per direct unit of measure}\right)} = \frac{\text{Estimated overhead cost}}{\text{Projected units of measure}} \tag{4.1}$$

For example, assume that the estimated overhead costs are $450,000 and management projects that 25,000 direct labor hours will be required during the period.

$$\underset{\text{(per direct labor hour)}}{\text{Predetermined overhead rate}} = \frac{\text{Estimated overhead cost}}{\text{Estimated direct labor hours}}$$

$$= \$450{,}000/25{,}000 \text{ hours}$$

$$= \$18 \text{ per direct labor hour}$$

Overhead costs are then applied to individual products on the basis of the number of labor-hours required to produce each unit. For example, if it takes 1/2 hour of direct labor to produce 1 unit of the product, that unit will be assigned $9 of manufacturing overhead cost. This amount would be added to the direct materials and direct labor costs to arrive at the overall manufacturing unit cost.

Example 4.1

Assume that, during a specific period, a company produced 6,000 units of its product. The production costs during the period included the following:

Raw Materials	$6,000
Direct Labor	10,800
Factory Overhead	5,400

The company can determine the production unit cost as follows:

Raw Materials ($6,000/6,000 units)	$1.00
Direct Labor ($10,800/6,000 units)	1.80
Factory Overhead ($5,400/6,000 units)	0.90
Unit Cost (overall)	$3.70

In this example, where all of the information is known, the unit cost was computed at the completion of the job. However, before completion of the job, such perfect information is not readily available. Then, the company needs a good estimate of the unit cost to use for price mark-ups. The following is an example of applying a predetermined overhead rate.

Example 4.2

Assume that the cost analyst of a company, based on historical cost information, has developed the following data for one of the company's products:

Raw Materials	$3.50 per unit
Direct Labor	$8.50 per unit
Factory Overhead	50% of direct labor cost

Based on the predetermined overhead rate of 50% of direct labor cost, the analyst may compute the unit cost as follows:

Raw Materials		\$3.50
Direct Labor		8.50
Factory Overhead	0.5 × \$8.50 =	4.25
Unit Cost		\$16.25

The computed unit cost of \$16.25 is merely an estimate based on a predetermined overhead rate and may not be an exact cost. Nevertheless, it is very helpful for product pricing and job costing purposes.

4.3.2. Overhead Accumulation Method

Another method of assigning overhead costs to products is to accumulate all manufacturing overheads in cost pools during a period. Then, allocate the accumulated amounts to the product lines. This method, which according to the process costing system (as opposed to the job-order costing system), involves no guesswork and results in a fairly accurate allocation of overhead costs to products manufactured. However, this method of costing has distinct disadvantages:

- It treats all products as being identical. Whereas, the products that require more materials and time than others should absorb more overhead costs.
- A firm cannot compute the cost of goods manufactured until the end of a period. Therefore, performing cost analysis is rather difficult during the period.
- Such a method does not facilitate an advance determination of a product's price under the cost-plus pricing method. To obtain an effective pricing tool, the manufacturer needs to know the cost prior to receiving an order to quote the selling price for a product without waiting for the completion of the ordered job.

In a traditional costing method, a firm may apply a single **plantwide allocation rate**; or they may apply a single overhead allocation rate for each department. The firm may use the **department allocation** approach, which applies several allocation rates (one for each department) and therefore uses several predetermined overhead rates.

4.4. ACTIVITY-BASED COSTING (ABC)

Under the traditional methods, the measures of indirect costs allocated to products or departments are somewhat arbitrary. However, accountants have developed activity-based costing (ABC), which is a more objective procedure for costs allocation. Activity-based costing provides more detailed measures of costs than the traditional methods. The Chartered Institute of Management Accountants (CIMA) defines ABC as:

> an approach to the costing and monitoring of activities which involves tracing resource consumption and costing final outputs. Resources are assigned to activities, and activities to cost objects based on consumption estimates. The latter utilize cost drivers to attach activity costs to outputs. (CIMA, 2008)

Activity-based costing (ABC) is a costing method that assigns overhead (indirect) costs to activities and to the cost objects (divisions or products) based on their use of activities. That is, ABC is based on the concept that activities consume resources (cost centers), and divisions or products (cost objects) consume activities. Figure 4.1 illustrates this concept for a hypothetical automobile manufacturing company.

Under ABC, a firm may use a single-stage or a two-stage process for allocating its overheads. A **single-stage** cost allocation process uses a single, plantwide, rate to allocate costs. Whereas, a **two-stage** system first allocates costs to departments or activities and then allocates costs from the departments or activities to the products or services. However, the two-stage provides more accurate results in allocating costs.

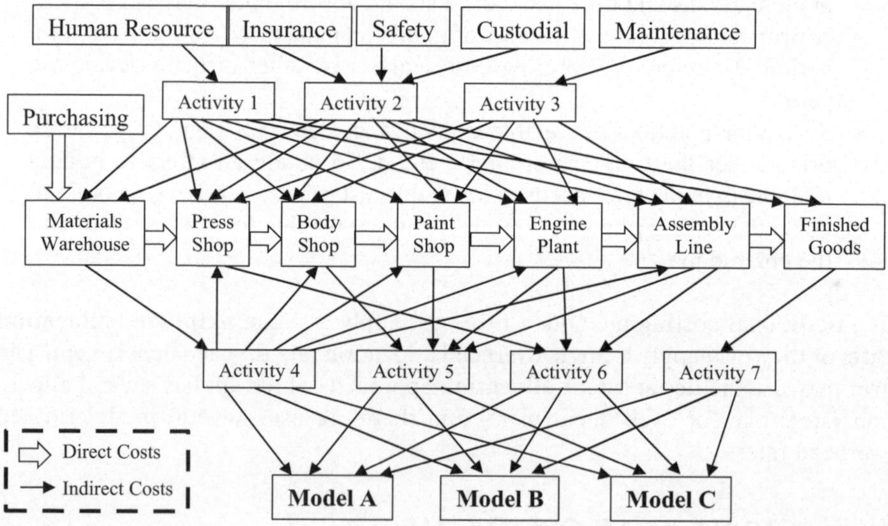

FIGURE 4.1 ABC cost centers (departments), activities, and cost objects (automobile models).

4.4.1. Bases for Overhead Allocation

Each department within a company incurs costs or expenses. Therefore, each department is a cost center. Overhead costs are allocated to the affected cost centers based on some output measures, known as cost drivers. A cost driver or allocation base is a basis upon which an overhead cost is allocated to the cost objects (production and service departments, products, jobs, or services). An allocation base takes a quantity form, such as machine hours used, kilowatt-hours consumed, or square-footage occupied. In many situations, a firm must allocate its overheads to several departments or cost centers within the factory. As examples, we present some practical bases for overhead costs allocation in the following subsection.

When the use of a resource or the performance of an activity changes, the level of the cost driver (or output measure) will also change, causing changes in costs. Figure 4.2 presents examples of cost drivers for various types of service costs (overheads).

A related concept to activity-based costing is Activity-Based Management (ABM). Through ABM, managers try to identify activities that consume resources. Then, they would develop strategies to manage costly activities to reduce costs and improve quality.

Cost Driver (Allocation Basis)	Service Cost (Overhead)
Number of employees	Factory lunchroom costs, factory dispensary medical services, human resource costs, and health insurance.
Direct materials,	Indirect materials.
Number of requisitions	Indirect materials, maintenance/repair and warehouse costs.
Number of machines	Machinery maintenance and repair costs.
Machine hours	Machinery maintenance and repair costs; and depreciation.
Number repair requests	Machinery maintenance.
Kilowatt-hours	Light and power (if meters are installed)
Floor space occupied	Building depreciation, and maintenance, lease/rent, fire insurance, property taxes, safety and security and custodial costs.
Investment in equipment	Machinery depreciation and fire insurance.
Direct labor cost	Payroll taxes and worker compensation insurance.
Maintenance labor hours	Maintenance and repair costs
Employee training hours	Human resource costs.
Units/volumes handled	Warehouse, shipping and receiving costs.
Capacity of machines	Light & power costs.
Number of computers	Information technology service costs.
Weights of laundry	Laundry costs
Number of flights	Airport ground service costs.

FIGURE 4.2 Examples of costs drivers for various service costs (overheads).

Example 4.3

Shell Tech Company allocates its indirect material costs to its cost centers (departments) on the basis of direct charges according to the store requisitions, as follows:

Distribution of indirect material costs	
Department X	$4,200
Department Y	3,500
Department Z	1,000
Maintenance Department	2,500
Warehouse	900
Total	**$12,100**

The company allocates indirect labor costs to the cost centers (departments) on the direct charges basis, according to the time card analysis, as follows:

Distribution of indirect labor costs	
Department X	$3,800
Department Y	3,000
Department Z	2,000
Maintenance Department	2,500
Warehouse	1,000
Total	**$12,300**

The total cost of light and power, which was $3,000, is allocated in accordance with the number of kilowatt-hours used. Therefore, the allocation rate will be 0.03 per kW-h (i.e., $3,000/100,000).

Distribution of light and power costs

	kW-h Used	Light and Power Cost
Department X	38,000	$1,140
Department Y	26,000	780
Department Z	20,000	600
Maintenance Department	11,000	330
Warehouse	5,000	150
Total	**100,000**	**$3,000**

Payroll taxes, amounting to $861, are allocated in accordance with the indirect labor costs charged to each department, as follows: ($861/$12,300 = $0.07) × (Indirect Labor Cost)

Distribution of payroll taxes

	Indirect Labor	Payroll Taxes
Department X	$3,800	$266
Department Y	3,000	210
Department Z	2,000	140
Maintenance Department	2,500	175
Warehouse	1,000	70
Total	**12,300**	**$861**

A total of $1,000 fuel was directly used by and charged to the Maintenance Department. The cost of repairs amounted to $2,840, was directly charged to each department, as follows:

Distribution of repair costs

Department X	$1,000
Department Y	670
Department Z	195
Maintenance Department	560
Warehouse	415
Total	**$2,840**

Rent and Fire Insurance costs were, $6,000 and $1,200, respectively, allocated in proportion to the floor space occupied (square footage), as follows:

Distribution of rent and insurance costs

	Floor Space	Ratio (%)	Rent	Fire Insurance
Department X	3,200	40	$2,400	$480
Department Y	2,400	30	1,800	360
Department Z	1,200	15	900	180
Maintenance Department	800	10	600	120
Warehouse	400	5	300	60
Total	**8,000**	**100**	**$6,000**	**$1,200**

The depreciation of machinery is allocated using the straight-line method based on cost, as follows:

Distribution of depreciation cost

	Cost of Machinery	Depreciation Charged
Department X	$25,000	$5,000
Department Y	20,000	4,000
Department Z	15,000	3,000
Maintenance Department	10,000	2,000
Warehouse	1,200	240
Total	**$71,200**	**$14,240**

Sundry (miscellaneous) costs, of total $1,850, are allocated in proportion to the number of employees in each department, as follows:

Distribution of sundry costs

	Number of Employees	Ratio (%)	Sundry Costs
Department X	40	40	$740
Department Y	20	20	370
Department Z	20	20	370
Maintenance Department	15	15	278
Warehouse	5	5	92
Total	**100**	**100**	**$1,850**

All allocated costs are recorded in a "worksheet for allocation of factory overhead." Then the overhead costs allocated to service departments are reallocated to operating/production departments. Assume that the company reallocates the costs of the Storeroom first, according to the number of requisitions filled, as shown below:

	Number of Requisitions	Ratio (%)
Department X	1,000	40
Department Y	750	30
Department Z	500	20
Maintenance Department	250	10
Total	**2,500**	**100**

The Maintenance Department's costs are reallocated next, according to floor space occupied by the production departments, as listed below:

Department	Floor Space (sq. ft.)	Ratio
Department X	3,200	8/17
Department Y	2,400	6/17
Department Z	1,200	3/17
Total	**6,800**	**1.00**

The entire overhead cost allocation for this example is summarized in the worksheet shown in Figure 4.3.

Shell Tech Company
Work-Sheet for Allocation of Factory Overhead
For the Month of July, 2006

			Production Departments			Service Dept.	
Overhead Cost Item	**Allocation Basis**	**Total**	X	Y	Z	Maint-enance	Store-room
Indirect Materials	Materials Used	$12,100	$4,200	$3,500	$1,000	$2,500	$ 900
Indirect Labor	Labor Performed	12,300	3,800	3,000	2,000	2,500	1,000
Payroll Taxes	Wages	861	266	210	140	175	70
Fuel	To Maintenance	1,000				1,000	
Light and Power	kW-h used	3,000	1,140	780	600	330	150
Repairs	Cost of Repairs	2,840	1,000	670	195	560	415
Rent	Floor Space	6,000	2,400	1,800	900	600	300
Depreciation, Machinery	Cost of Machine	14,240	5,000	4,000	3,000	2,000	240
Fire Insurance	Floor Space	1,200	480	360	180	120	60
Sundry Costs	No. of Employees	1,850	740	370	370	278	92
TOTALS		$55,391	$19,026	$14,690	$8,385	$10,063	$3,227
Reallocation of Service Department Costs: Storeroom Costs	No. of Requisitions	--------	1,291	968	645	323	(3,227)
		$55,391	$20,317	$15,658	$9,030	$10,386	---
Maintenance Dept. Costs	Floor Space	--------	4,888	3,666	1,832	(10,386)	
Final Cost Distribution		**$55,391**	**$25,205**	**$19,324**	**$10,862**	**---**	**---**

FIGURE 4.3 A worksheet for overhead cost allocation for Example 4.3.

4.4.2. Overhead Cost Pools

Cost analysts and accountants follow a systematic process to allocate all costs to the produced units. To accomplish this, they use cost pools for allocating their factory overheads to units of products they produce. A ***cost pool*** is a grouping of all overhead costs associated with performing a particular task, such as maintenance, warehousing, human resources, security and safety, and physical plant services. Accountants allocate costs in the cost pools to the operating and support departments based on the cost drivers (activities) and, then, they reallocate the costs assigned to the support departments to the operating departments. For example, a firm accumulates the costs of the Maintenance Department in a cost pool and, then, allocates the costs in this pool to the departments that have used its services. The difference between operating (or production) and support (or service) departments is explained as follows:

- **Operating/Production Departments**: An operating/production department is directly involved with manufacturing products; for instance, each of the several stations in an assembly line is an operating department.
- **Support Departments**: A support/service department is not directly involved with producing products, but it renders services to the operating departments; examples of support departments are: warehouse, building maintenance, human resources department, custodial services, purchasing department, and transportation department.

Activity-Based Costing (ABC) uses cost pools to allocate costs to its activities. An activity is the smallest possible unit of work. An activity consumes resources, such as staff and machine times or funds. Conceptually, an activity is any action that converts an input into an output.

4.5. METHODS FOR REALLOCATING COSTS OF SUPPORT DEPARTMENTS

The costs allocated to the support departments are the company's overhead costs. Thus, they will be reallocated to the operating departments. There are three popular methods for reallocating service department costs to operating departments: ***direct*** method, ***step-down*** method, and ***reciprocal*** method. It is important to realize that regardless of the method used to allocate support department costs, total factory overhead costs remain unchanged. However, the different allocation methods split up the costs differently among the production departments.

4.5.1. Direct Method of Reallocating Support Costs

The direct method is the simplest method of allocating the costs of service departments to production departments. Under this method, a firm does not allocate the costs incurred by a service department to any other service departments; it rather directly allocates them to the production departments based on relative uses.

Suppose, for example, a firm has two production and two service departments; whereas departments P1 and P2 are production departments, and departments S1 and S2 are service departments. Under the direct method of cost allocation, the firm allocates the costs incurred by departments S1 and S2 to departments P1 and P2 only. It would not allocate the costs of departments S1 and S2 to each other even though these two departments provide service to each other.

Example 4.4

The Lantar Company has two service departments and two operating departments. The costs incurred during a given per are summaries as follows:

	Production Departments		Service Departments		
	P1	**P2**	**S1**	**S2**	**Total**
Departmental costs ($)	255,550	335,850	99,000	68,400	758,800
Labor hours	10,000	14,000	3,000	8,000	29,000
Floor space occupied (sq. ft.)	13,000	27,000	12,000	8,000	65,000

The two service departments provide services to production departments and to each other as well. The company allocates the costs of department S1 based on direct labor hours and the costs of department S2 based on floor space occupied. Determine how the company would allocate the costs of the service departments to the production departments using the direct method of cost allocation.

Solution

The allocation of department S1's costs based on the labor-hour rate:

Rate = $99,000/(10,000 + 14,000) = $4.125 per labor hour
Cost allocated to department P1: $4.125(10,000) = $41,250
Cost allocated to department P2: $3.3875(13,450) = $57,750

The allocation of department S1's costs based on the labor-hour proportions

Proportion to production department P1 = 10,000/(10,000 + 14,000) = 5/12
Proportion to production department P2 = 14,000/(10,000 + 14,000) = 7/12
Cost allocated to department P1: (5/12)(99,000) = $41,250
Cost allocated to department P2: (7/12)(99,000) = $57,750

The allocation of department S2's costs based on per square foot floor space rate:

Rate = $68,400/(13,000 + 27,000) = $1.71 per sq. ft.
To department P1: $1.71(13,000) = $22,230
To department P2: $1.71(27,000) = $46,170

The allocation of department S2's costs based on floor space proportions:

Portion to production department P1 = 13,000/(13,000 + 27,000) = 0.325 (or 32.5%)
Portion to production department P2 = 27,000/(13,000 + 27,000) = 0.675 (or 67.5%)
Cost allocated to department P1: (0.325)(68,400) = $22,230
Cost allocated to department P2: (0.675)(68,400) = $46,170

The following is a summarized cost allocation using the direct method:

	Production Departments		Service Departments		
	P1	**P2**	**S1**	**S2**	**Total**
Cost before allocation	$255,550	$335,850	$99,000	$68,400	$758,800
Allocation of:					
Department S1's costs	$41,250	$57,750	(99,000)		0
Department S2's costs	$22,230	$46,170		(68,400)	0
Cost after allocation	**$319,030**	**$439,770**	**$0**	**$0**	**$758,800**

The major advantage of the direct method for allocating service department costs is its simplicity and ease of use. However, it comes with two major disadvantages. First, it ignores the mutual services among service departments. This may hinder reviewing the actual costs of such a department for finding cost reduction potentials. The second disadvantage tails into the first one, and the inaccuracy in the actual service costs can distort the final cost allocations to the products.

Periodically, cost accountants prepare a worksheet to allocate the factory overhead costs to various cost centers (product lines and service departments).

4.5.2. Step-Down Method of Reallocating Support Costs

The step-down (or ***step*** method or ***sequential***) method allocates the costs of a service department to other service departments as well as to production departments. Under the step method, the cost allocation is a sequential process. It starts with allocating the costs of the service department that provides the highest level of service to other service departments to production departments and all other service departments. It ends with allocating the costs of the last service department (including the absorbed costs from other service departments), which provides the least amount of service to other service departments, to the production departments. This method follows the following steps:

1) Select a service/support department and allocate its costs to the production departments and support departments to which it provides services. Give priority to the support department that provides the highest level of service to all service departments. If this is not possible, then choose the service

department with the greatest total costs. Do not allocate costs back to the service department itself.

2) Select another service department and allocate its cost to the production departments and the remaining support departments. Do not allocate any cost back to the service department whose costs have already been allocated.
3) Proceed in this manner until the costs of all the service departments have been allocated to the production departments.

The following is an example of allocating (overhead) costs of support/service departments using the step-down (or sequential) method.

Example 4.5

Recall the cost information for the Lantar Company, in the **previous example**, as restated below. Allocates the costs of department S1 based on direct labor hours and the costs of department S2 based on floor space occupied using the **step-down** method.

	Production Departments		Service Departments		
	P1	**P2**	**S1**	**S2**	**Total**
Departmental costs ($)	255,550	335,850	99,000	68,400	758,800
Labor hours	10,000	14,000	3,000	8,000	29,000
Floor space occupied (sq. ft.)	13,000	27,000	12,000	8,000	65,000

Solution

We need to determine the costs of the service department of which allocation is done first. We can determine this as follows:

Total labor hours of service provided by department S1:
 10,000 + 14,000 + 8,000 = 32,000 sq. ft.
Proportion of service of department S1 provided to S2:
 8,000/32,000 = 25%
Total floor space served by department S2:
 13,000 + 27,000 + 10,000 = 50,000 sq. ft.
Proportion of service of department S2 provided to S1:
 10,000/50,000 = 20%

Since department S1 provided a higher proportion of its service to department S2 than vice versa, we start with allocating the costs of S1 to other departments. Then, we allocate the updated costs of the service department S2 to the production departments.

The allocation of department S1's costs based on labor-hour proportions

Total labor hours used to serve other departments by S2 = 10,000 + 14,000 + 8,000
= 32,000 hours
Proportion to service department S2 = 8,000/32,000 = 1/4 = 25%
Proportion to production department P1 = 10,000/32,000 = 5/16 = 31.25%
Proportion to production department P2 = 14,000/32,000 = 7/16 = 43.75%
Cost allocated to department S2: (1/4)(99,000) = $24,750
Cost allocated to department P1: (5/16)(99,000) = $30,938 (rounded up)
Cost allocated to department P2: (7/16)(99,000) = $43,312 (rounded down)

The allocation of department S2's costs based on per square foot floor space rate:

Total cost of department S2 to allocate to departments P1 and P2:
$68,400 + 24,750 = $93,150
Rate = $93,150/(13,000 + 27,000) = $2.32875 per sq. ft.
To department P1: $2.32875(13,000) = $30,274
To department P2: $2.32875(27,000) = $62,876

The following is a summarized cost allocation using the step-down method:

	Production Departments		Service Departments		
	P1	**P2**	**S1**	**S2**	**Total**
Cost before allocation	$255,550	$335,850	$99,000	$68,400	$758,800
Allocation of:					
Department S1's costs	$30,938	$43,312	(99,000)	24,750	0
Department S2's costs	$30,274	$62,876		(93,150)	0
Cost after allocation	**$316,762**	**$442,038**	**$0**	**$0**	**$758,800**

4.5.3. Reciprocal Method of Reallocating Support Costs

The reciprocal method fully recognizes the reciprocal services provided by support departments to other support departments. As a result, the reciprocal is the most accurate cost allocation method among the three. This method uses simultaneous equations to allocate the costs incurred by each service department to other departments. The total cost of a support department is the sum of **its own direct costs** and the **allocated costs of other support departments**.

Example 4.6

Recall the cost information for the Lantar Company, in the **previous example**, as restated below. Allocates the costs of department S1 based on direct labor hours and the costs of department S2 based on floor space occupied using the **reciprocal** method.

	Production Departments		Service Departments		
	P1	P2	S1	S2	Total
Departmental Costs ($)	255,550	335,850	99,000	68,400	758,800
Labor Hours	10,000	14,000	3,000	8,000	29,000
Floor Space Occupied (sq. ft.)	13,000	27,000	12,000	8,000	65,000

Solution

As we did under the step-down method, we need to determine the proportion of the services received by each service department from other service departments, as restated below:

Total labor hours of service provided by department S1:
10,000 + 14,000 + 8,000 = 32,000 sq. ft.
Proportion of service of department S1 provided to S2:
8,000/32,000 = 25% of S1 costs to S2
Total floor space served by department S2:
13,000 + 27,000 + 12,000 = 52,000 sq. ft.
Proportion of service of department S2 provided to S1:
12,000/52,000 = 3/13 (23.077%) of S2 costs to S1

The reciprocal method allocates 25% of the total cost of service department S1 to service department S2; and 20% of the total cost of department S2 to department S1, as expressed by the following cost equations:

S1 – $99,000 + (3/13)S2
S2 = S68,400 + 0.25S1

By the substitution method, the costs of department S1 (designated by S1) becomes:

S1 = 99,000 + (3/13)($68,400 + 0.25S1)
(49/52)S1 = $114,785.6154
S1 = $121,812 (equals departmental cost of $99,000 *plus* $22,812 service cost from S2)

The costs of department S2 (designated by S2) becomes:

S2 = $68,400,000 + 0.25[$99,000 + (3/13)S1]
(49/52)S2 = $93,150
S2 = $98,853 (equals departmental cost *plus* $30,453 service cost from S1)

Note that the sum of the costs of departments S1 and S2 is higher than the sum of their direct costs.

Now, we allocate the costs of the service departments to the production department. We use the direct method but not the step-down method as we have already performed the mutual service costs allocation between the two service departments.
The allocation of total costs of department S1 to production departments based on labor hours in all other departments, including the service department S2:

Total labor hours to account for = 10,000 + 14,000 + 8,000 = 32,000 hours
Proportion to production department P1 = 10,000/32,000 = 5/16
Proportion to production department P2 = 14,000/32,000 = 7/16
Cost allocated to department P1: (5/16)($121,812) = $38,066 (rounded)
Cost allocated to department P2: (7/16)($121,812) = $53,293 (rounded)

To allocate the cost of department S2 based on floor space occupied by all other departments, including the service department S1:

Total floor space served by S2 = 13,000 + 27,000 + 12,000 = 52,000
Proportion to production department P1 = 13,000/52,000 = 13/52
Proportion to production department P2 = 27,000/52,000 = 27/52
To department P1: (13/52)($98,853) = $24,713
To department P2: (27/52)($98,853) = $51,328

The following is a summarized cost allocation using the direct method:

	Production Departments		Service Departments		
	P1	**P2**	**S1**	**S2**	**Total**
Cost before allocation	$255,550	$335,850	$99,000	$68,400	$758,800
Allocated cost of:					
Department S1	38,066	53,293	(121,812)	30,453	0
Department S2	24,713	51,328	22,812	(98,853)	0
Cost after allocation	**$318,329**	**$440,471**	**$0**	**$0**	**$758,800**

After allocating the service costs, the firm can divide the final costs of the production departments by their corresponding output units to determine the manufacturing unit cost for each product.

SUMMARY

Cost allocation is the process of assigning common overhead costs to two or more cost objects (e.g., products, divisions/departments, customers). Ideally, cost allocation must reflect the cause-and-effect relationship between cost pools and the cost objects to which they are allocated.

Firms try to select the basis for cost allocation that is the underlying cost driver. For example, a firm may choose direct labor hours as the cost driver for allocating its indirect costs (overheads). It would divide the total overhead costs incurred in a fiscal period by the total number of direct labor hours for the period. Then, it applies the per-hour overhead rate to each cost object based on the number of direct labor hours used by the cost object.

Under the traditional allocation method, the firm chooses one overhead rate as a means to allocate its overhead costs. Activity-based costing (ABC) is a more complex but accurate approach to allocating overheads. Instead of using a single factor in traditional methods, the ABC uses several factors to determine how to allocate various overheads. Each factor is tied directly to an aspect of overheads. Then, the firm allocates cost to each product based on how much of each element of overheads is used to produce the product.

The ABC allocates overheads in two stages. In the first stage, a firm determines the overhead cost per unit of each activity that causes (drive) costs. For example, say that the total cost of the human resources (HR) department is $300,000 per year and the firm has 500 employees. The first stage of allocation would be to compute the HR's cost per employee is $600. The second stage is to allocate the activity cost to each department based on the number of employees. For this example, if department X consists of 75 employees, $45,000 of HR's annual costs (i.e., $600 × 75 employees) is assigned to department X. The use of ABC can be summarized as follows:

1) **List Cost Pools (Cost Centers)**: Identify and categorize all resources as cost pools or cost centers. Determine the amount of the costs pertaining to each cost pool during the fiscal period.
2) **List Cost Objects**: List the relevant cost objects that consume resources (i.e., departments, product lines, customers).
3) **List Cost Drivers (Activities)**: activities involved in the manufacturing process link the cost pools to the cost objects. For example, labor hours, machine hours, set-up frequencies, number of machines, number of employees, floor spaces occupied, and testing.

Service department cost allocations are required to include the costs of support services in the costs of products and services provided. The three major methods for allocating the service department costs are the direct method, the step method, and the reciprocal method. The use of the methods depends on the extent to which services are provided by one service department to another.

Traditional costing simply applies an average (single) overhead rate to add overhead costs to the direct costs of manufacturing products, which is best used when the overhead of a company is low relative to the direct costs of production. Activity-based costing identifies all of the operations that are indirectly related to the manufacture of each product or department.

Process costing is used for mass production (i.e., when high volumes of a standardized product is produced) according to the following five steps.

Step 1: Cost Accumulation. Lists all costs in the beginning WIP and add the costs added during the period to get "Total Costs to Account For."

Step 2: Physical Flow of Units. Determine the total number of units worked during the period of interest to get "**Total Units to Be Accounted For**";

and find what happened to those units to get "Total Units Accounted For." These are computed as follows:

Beginning work in process units + units started this period
= **Total Units to Be Accounted For**
Units completed and transferred + Ending work in process units
= **Total Units Accounted For**

Step 3: Equivalent Units. Equivalent units are the number of units completed and transferred out plus the equivalent number of complete units for the partially completed units in the WIP.

Step 4: Cost per Equivalent Unit. Compute the total costs incurred during the period and the cost of the beginning WIP, and divide by the Total Equivalent Units calculated in Step 3. The formula is applied to direct materials, direct labor, and overhead individually. The calculation is:

$$\text{Cost/Unit} = \frac{(\text{Cost of beginning work in process inventory}) + (\text{costs added this period})}{(\text{Total equivalent units})}.$$

Step 5: Assign and Reconcile Costs. Multiply the equivalent units in WIP calculated in Step 3 by the cost per equivalent unit calculated in Step 4 to get the cost of ending WIP inventory. Multiply units completed and transferred by the cost per unit calculated in Step 4 to get the cost units completed and transferred out (in the final stage, the result will be the Cost of Goods Manufactured).

REFERENCE

CIMA, (2008). *CIMA Official Terminology.* The Chartered Institute of Management Accountants (CIMA). Retrieved January 21, 2021 from: http://www.cimaglobal.com/Documents/ImportedDocuments/cid_tg_activity_based_costing_nov08.pdf.pdf.

REVIEW QUESTIONS

R4.1. Explain why companies allocate costs. What is the main issue in cost allocation?
R4.2. What are the different methods of cost allocation?
R4.3. What does it mean by cost allocation?
R4.4. What is the purpose of cost allocation?
R4.5. What are the steps in overhead allocation using predetermined rates?
R4.6. What are the three common methods of allocating costs of support/service departments to operating departments?
R4.7. What is the difference between a service department and a production/operating department?

R4.8. What does the term "reciprocal services" refer to?

R4.9. Explain how direct, step-down, and reciprocal methods allocate the costs of service departments to the production departments.

R4.10. Explain how a sequence of service departments is chosen under the step-down method for allocating the costs of service departments to the production departments.

R4.11. Explain the "single-stage" and "two-stage" cost allocation processes.

R4.12. Describe the plantwide and department rates for overhead cost allocation.

R4.13. What are the differences and similarities among the direct, step-down, and reciprocal methods of allocating costs of service departments?

R4.14. What criterion should be used to determine the order of reallocating the costs of service departments under the step-down method?

PROBLEMS

P4.1. The following worksheet for allocation of factory overhead of the Waldoor Company for the month of August 20XX is given. Assume that the Warehouse has received less service from the Building Service Department than the other way around. So, first reallocate costs of the Warehouse to the Building Department, according to the number of store requisitions filled. Then, reallocate costs of the Building Service Department according to the floor space occupied.

Waldoor Company
Worksheet for Allocation of Factory Overhead
For the Month of August, 20XX

			Production Department			Service Department	
Overhead Cost Item	**Allocation Basis**	**Total**	**A**	**B**	**C**	**Building**	**Warehouse**
Indirect Materials	Materials Used	$21,425	$6,045	$5,400	$3,350	$4,915	$1,715
Indirect Labor	Labor Performed	27,000	6,750	5,000	4,000	3,750	7,500
Payroll Taxes	Wages	810					
Light and Power	kW-h Used	3,500					
Repairs	Cost of Repairs	2,740	1,250	430	250	635	175
Depreciation, Machinery	Cost of Machine	7,500	2,640	1,800	1,200	1,860	—
Fire Insurance	Floor Space	1,300					
Other Expenses	No. of Employees	1,800					
TOTAL		$66,075					
Reallocation of Service Department Costs:							
Warehouse Costs	No. of Requisitions						
Building Service Department Costs	Floor Space						
Final Cost							

Required: Based on the following information, complete the worksheet.

Department	Light and Power (kW-h)	Floor Space (sq. ft.)	No. of Employees	No. of Store Requisition
Department A	35,000	3,500	40	40
Department B	30,000	2,500	25	20
Department C	22,000	2,000	20	12
Building Service	9,000	1,200	10	8
Warehouse	4,000	800	5	—
Totals	**100,000**	**10,000**	**100**	**80**

P4.2. Delta Company has two service departments, Maintenance (Maint) and Warehouse (Ware), and two operating departments (Departments A and B). The costs of the Maintenance Department are allocated based on the floor space occupied. The costs of Warehouse are allocated based on the number of store requisitions. The following information is provided:

	Service Department		Operating Department	
	Maint	Ware	A	B
Costs ($)	90,000	45,000	180,000	250,000
Floor Space (sq. ft.)	10,000	20,000	30,000	50,000
Number of Requisitions	7,000	500	10,000	15,000

Required: Allocate the costs service departments using:

a) The direct method.

b) The step-down method.

c) The reciprocal method.

P4.3. Using the provided worksheet allocate MKATZ Company's factory overheads to its service departments (Maintenance and Warehouse) and production departments (X, Y, and Z).

MKATZ Company
Worksheet for Allocation of Factory Overhead
For the Month of July, 20XX

			Production Departments			Service Department	
Overhead Cost Item	**Allocation Basis**	**Total**	**X**	**Y**	**Z**	**Maintenance**	**Warehouse**
Indirect Materials	Materials Used	$24,300					
Indirect Labor	Labor Performed	17,400					
Payroll Taxes	Wages	2,460					
Fuel	To Maintenance	1,134					
Light and Power	kW-h Used	3,400					
Repairs	Cost of Repairs	2,840					
Rent	Floor Space	6,000					
Depreciation, Machinery	Cost of Machine	14,250					
Fire Insurance	Floor Space	1,200					
Sundry Costs	No. of Employees	1,800					
TOTALS		$57,484					
Reallocation of Service Department Costs:							
Warehouse Costs	No. of Requisitions						
Maintenance Department Costs	Floor Space						
Final Cost Distribution							

A worksheet for overhead cost allocation for Problem P4.4.
The following data were obtained from the company's records:

Department	Number of Employees	Floor Space (sq. ft.)	Number of Requisitions	kW-hrs Used
Production X	60	5,500	1,000	33,000
Production Y	50	4,200	800	24,600
Production Z	34	3,800	700	16,900
Maintenance	10	2,000	1,500	3,200
Warehouse	6	4,500	—	2,300
Total	**160**	**20,000**	**4,000**	**80,000**

- Indirect materials are charged as they are requested by the consuming department, the indirect labor is allocated to the cost centers according to time card recordings, and repair costs are also charged directly where repairs occurred, as follows:

Department	Indirect Materials	Indirect Labor	Repair Cost
Production X	$8,500	$4,600	$998
Production Y	6,500	4,200	670
Production Z	4,000	4,300	490
Maintenance	20,500	30,500	357
Warehouse	—	12,900	345
Total	**$39,500**	**$56,500**	**$2,860**

- Payroll taxes, amounting to $11,280, are allocated based on the indirect labor cost charged to each department.
- A total of $3,600 fuel cost was directly used by and charged to the Maintenance Department.
- The light and power cost of $3,200 is allocated according to the number of kilowatt-hours used.
- Rent and Fire Insurance costs were, $9,000 and $1,200, respectively, are based on the floor space occupied by each department.
- Depreciations of machinery were computed according to the company's depreciation schedules and are as shown below:

Department	Depreciation Charged
Production X	$2,200
Production Y	1,800
Production Z	1,200
Maintenance	1,000
Warehouse	500
Total	**$6,700**

- The sundry (miscellaneous) expenses of a total of $320 are allocated based on the number of employees in each department, as follows:
- The overhead allocated to service departments is reallocated to production departments using the "step-down" method, assuming that the Warehouse has provided more services to the Maintenance Department than receiving from it. The Warehouse costs are reallocated based on the number of store requisitions and the Maintenance Department's costs are reallocated based on the floor space occupied.

P4.4. The Talesh Company has the following data pertaining to its three services and two operating departments:

	Service Departments			Operating Departments		
	S1	**S2**	**S3**	**A**	**B**	**Total**
Overhead Costs	$84,000	$66,000	$40,000	$90,000	$97,000	$377,000
Number of Employees	15	25	15	110	90	255
Space Occupied (sq. ft.)	10,000	15,000	20,000	100,000	80,000	255,000
Machine Time (hours)	—	—	—	15,000	25,000	40,000

The company allocates service department costs as follows:

1. Allocates Department S1's costs based on the number of employees.
2. Allocates Department S2's costs based on the square footage of space occupied.
3. Allocates Department S3's costs based on the hours of machine time.

Required

Allocate the cost of service departments to operating departments using the step-down method in descending order of their costs.

5 Joint Cost Allocation

LEARNING OBJECTIVES

- Describe joint products and joint costs.
- Identify the split-off point(s) in a common process for multiple products.
- Distinguish joint (main) products from by-products.
- State reasons for joint cost allocation.
- Allocate joint costs using various methods.
- Allocating cost to by-products.
- Decide to sell products at the split-off point or process further.
- Decide which products to produce.
- Apply mathematical programming to joint cost allocation.

5.1. INTRODUCTION

When processing a common input (material) results in producing two or more outputs (products or services), the output products are referred to as joint products (occasionally, as common products). Joint costs allocation is the process of assigning the common costs to the multiple products manufactured from a common input or process. Companies allocate costs to establish a cost basis for pricing or performance evaluation.

Figure 5.1 illustrates multiple products that are produced from the same set of inputs through the same manufacturing process. As shown in the figure, in such a manufacturing process, there is a point, called the "***split-off point***," before which it is not possible to identify the individual products. Beyond this point all products are clearly identifiable, and separated and sent through their corresponding special processing lines (if needed to complete the products). The costs incurred up to the split-off point are referred to as "***joint costs***" and any costs incurred beyond the split-off point (i.e., during the special process) are called "***separable costs***."

The following are three examples of multiple products that are jointly produced from the same input materials:

- Processing crude oil results in several different products, such as gasoline, diesel fuel, oil, kerosene, plastic, waxes, asphalt, and other oil-based products.
- Processing a cow, in meatpacking, results in joint products, such as rawhide, ground beef, steaks, and other meat products.
- Processing raw milk results in joint products, such as whole milk, skim milk, butter, cheese, cream, and other dairy products.

DOI: 10.1201/9780429432163-5

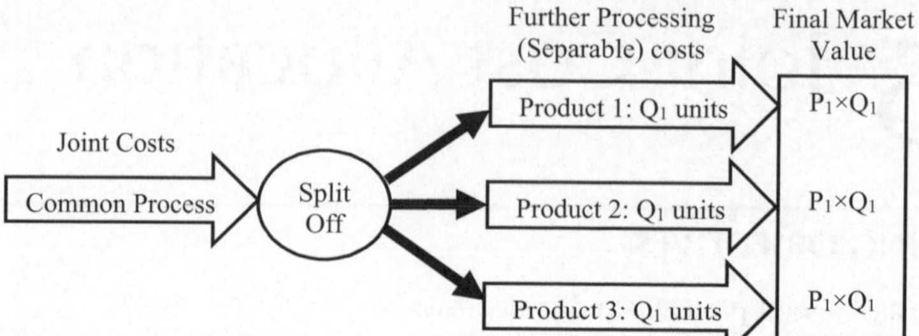

FIGURE 5.1 An illustration of production flow for manufacturing joint products through a common (or joint) process.

5.2. JOINT COST ALLOCATION TERMINOLOGY

The reader should be familiar with the terminologies used in cost accounting for multiple products in order to comprehend its concepts. In general, any of the manufactured multiple products can be a joint product, a common product, or by-product. The classification of multiple products depends on how they are produced and their market value. Joint and common products are defined as follows (Williams and Kennedy, 1983; Tayyari and Parsaei, 1992).

- ***Joint products*** are those multiple products of a single production process that are manufactured in ***fixed proportions***.
- ***Common products*** are those products that are manufactured in common by processing the same input materials and production facility in ***variable proportion***.
- A ***by-product*** is an output of a manufacturing process that adds a relatively small amount to the total market value of all outputs and considered as incidental product. For example, sawdust or wood barks are by-products of the lumber industry and feathers are by-products of poultry processing.

The distinction between main (joint and common) products and by-products is usually based on the market value of the product. A main (joint or common) product adds significantly to the total market value of all outputs. By-products do not add significantly to the total market value of all multiple products. By-products may also be defined as those joint products that are of minor importance, either because their production is incidental to the main objectives of the manufacturing process or because they have relatively little value. Hence, classification of a product as a main product or a by-product requires personal judgment.

Other important terms used in joint cost allocations include:

- **Main products** are products that a manufacturer intends to produce and have significant sales values as compared to the sales values of other product(s).

- **Intermediate products** are products that require further processing before becoming salable.
- **Final products** are products that are completed and ready for sale without requiring further processing.
- **Split-off point (SOP)** is the point at which a joint process ends and the individual products are identified and separated from each other, beyond which separate processes for the individual products begin if needed. This point of production may also be referred to as the ***point of recovery***.
- **Joint costs (JC)** include all costs of processing joint products prior to the split-off point.
- **Separable costs** are the costs incurred after the split-off point to complete the individual joint product.
- **Relative market value (RMV)** or **net-realizable value (NRV)**, in joint cost allocation, is an approximated sales value of a joint product at the split-off point, which equals the final market (sales) value less the needed further processing (separable) cost, as expressed in the following formula:

$$\text{RMV} = (\text{Selling Price} - \text{Unit Separable Cost}) \times (\text{Units Produced})$$

- **Scraps** are products with no values or negative sales values if the manufacturer must haul them away to a landfill.

5.3. REASONS FOR JOINT COST ALLOCATION

Joint costs are production costs. There are several important reasons for allocating joint costs to the individual products, which include the following:

- **Financial reports**: Joint costs are allocated to products to determine inventory values of the unsold units of the products and the cost of goods sold for both internal and external reports.
- **Product pricing**: The price of a product is usually set based on its total cost; whereas the joint costs are part of the total manufacturing cost.
- **Profitability analysis for decision-making**: Profitability analysis is used to determine which product to produce. Such analysis is based upon the manufacturing costs and sales prices of the products.
- **Decision about selling products at the split-off point**: The management may decide to sell some or all joint products at the split-off point when the further processing (separable) costs cannot be justified.
- **Contract reimbursement**: Cost-based contracts require justification and document for the incurred costs to authorize reimbursement, especially for government contracts.
- **Regulated products**: For some products, the prices are set and controlled by regulatory agencies. Prices are set based on the costs to the producers of the products.

- **Insurance settlement**: When an insurance claim is filed for damaged assets, documentation for each asset's value is required.
- **Litigations**: When a company initiates legal action or defends itself against a legal action regarding a product or service, it needs reasonable documentation, including cost documentation.

5.4. PROFITABILITY ANALYSIS AND DECISION SCENARIOS

The objective of producing products is usually to maximize profits. To provide a good basis for making decision about producing the products that maximize overall profit, the following analysis procedure becomes necessary:

1) Identify final products from the joint process.
2) Forecast the sales volume and price relationship for each product.
3) Estimate the further processing costs needed beyond the split-off point for converting each product into a final (or salable) product.
4) Continue producing the products that result in the overall maximum profit.

However, in the presence of multiple constraints and requirements, finding an optimal solution may require mathematical programming (e.g., linear programming and nonlinear programming). A firm that produces joint products may deal with the following decision scenarios:

1) Which product to produce?
2) Sell a product at split-off or process further?
3) How much of each product to produce?

5.4.1. Decision Scenario 1: Which Product to Produce?

When there are no constraints or requirements, it will be obvious the product(s) that generate(s) the maximum overall profit would be produced.

5.4.2. Decision Scenario 2: Should Sell a Product at Split-Off or Process Further?

A decision-maker may have to decide whether to sell a product at the split-off or process it further. In such a case, a product should be processed beyond the split-off point only if the **incremental revenue** exceeds its **incremental processing costs**.

In a "sell-or-process-further" decision, joint costs are sunk costs and irrelevant. Joint products have been produced, and a prospective decision must be made to take one of the following actions:

- Sell a product "as is" at the split-off point;
- Process further, which requires additional separable costs, and, then, sell at a higher price.

Example 5.1

A lumber yard cuts logs of wood from which unfinished lumber and wood chips are the joint products. The unfinished lumber can be sold "as is" or processed further into finished lumber.

Wood chips can also be sold "as is" for landscaping or processed further into composite boards. The data about the two joint products include:

	Per Log	
	Lumber	**Wood Chips**
Sales value at the split-off point	$1,200	$300
Sales value after further processing	2,600	350
Allocated joint costs	290	60
Cost of further processing	500	65

An analysis of selling at the split-off point or further process:

	Per Log	
	Lumber	**Wood Chips**
Sales value after further processing	$2,600	$350
Sales value at the split-off point	1,200	300
Incremental revenue	$1,400	$50
Cost of further processing	500	65
Incremental profit (loss) from further processing	$900	$(15)

Questions

a) Should they process the lumber further and sell the wood chips "as is" at the split-off point?
b) How about wood chips?

Answers

a) They should process lumber further as it generates an additional $900 profit.
b) They should not process wood chips further and sell them at the SOP "as is" since its further processing reduces the profit by $15 per log.

5.4.3. Decision Scenario 3: How Much of Each Product to Produce?

This scenario usually arises when the decision is made in the presence of one or more constraints or requirements. Examples of such constraints are the limitations on resources (e.g., materials, labor time, machine time availability) and the rate of consumption of the resources by the individual products. In such cases, mathematical programming models play a major role in determining a product-mix that maximizes the overall profit.

5.5. TRADITIONAL METHODS FOR JOINT COST ALLOCATION TO MAIN PRODUCTS

Joint costs are indivisible because the products cannot be obtained separately. All joint costs that are incurred prior to the split-off point must be allocated among the multiple products to determine a unit cost for each product. The unit costs are dependent upon the allocation scheme employed. Any costs incurred beyond the split-off point are separable and are directly traceable to a particular joint product thus they are directly assigned to that particular output. Several traditional methods for allocating joint costs to the multiple products are presented in many accounting books (e.g., Myer, 1972; Needles et al., 1993; Kaplan, 1982).

The following are the common methods for allocating joint costs to the joint (main) products:

- Physical-measure (or simple-average-cost or average-unit-cost) method.
- Weighted-average-cost method (an extension of physical-measure method).
- Sales-value method.
- Relative-market-value (or net-realizable-value) method.
- Constant-gross-margin method.

The following two common methods are used to account for by-products:

- Zero-cost method.
- Net-realizable-value (or reversal-cost) method.

5.5.1. Physical-Measure Method of Joint Cost Allocation

The physical-measure method (also known as **quantitative unit**, **physical quantity**, **simple-average-cost**, and **average-unit-cost** method) allocates joint costs to products in proportion to a physical measure (for example, volume or weight). The is the simplest method for allocating a joint production cost to joint products because it assumes all the joint products are of the same units and have the same market values. The average unit cost is calculated by dividing the total joint cost by the total number of units of joint products produced.

Example 5.2

AAA Chemical Company processes a single material into two separate products, X and Y. Both products are salable immediately at the split-off point. Product X sells for $15 per gallon and Product Y for $3 per gallon. During the month of June, the company incurred $16,000 for materials, labor, and overhead in processing 4,000 gallons of an input material. The process yielded 1,200 gallons of X and 2,000 gallons of Y. (Note that the total number of units of the output does not have to be the same as the number of units of the input.) The company sold 900 gallons of Product X and 1,500 gallons of Product Y during June. You are asked to allocate the $16,000 joint costs to Products X and Y using the physical-measure method.

Solution

You can solve the problem by either of the following two procedures:

First procedure for joint cost allocation under the physical-measure method:

Step 1: Computing the average unit cost:

$$\text{Average unit cost} = (\text{Total cost})/(\text{Total number of units of output})$$

$$= \$16,000/(1,200 + 2,000) = \$5.00 \text{ per gallon}$$

Step 2: Allocating the joint cost:

$$\text{Joint cost allocated to a Product} = (\text{Average unit cost}) \times (\text{Number of units})$$

$$\text{Joint cost allocated to Product X} = \$5.00 \times 1,200 \text{ gal.} = \$6,000$$

$$\text{Joint cost allocated to Product Y} = \$5.00 \times 2,000 \text{ gal.} = \$10,000$$

Second procedure for joint cost allocation under the physical-measure method:

Step 1: Computing the proportion of production outputs:

$$\text{Product production proportion} = (\text{Product output quantity})/(\text{Total output quantity})$$

$$\text{Total output quantity} = 1,200 + 2,000 = 3,200 \text{ gallons}$$

$$\text{Proportion of Product X output} = 1,200/3,200 = 0.375 \text{ (or 37.5\%)}$$

$$\text{Proportion of Product Y output} = 2,000/3,200 = 0.625 \text{ (or 62.5\%)}$$

Step 2: Allocating the joint cost:

$$\text{Joint cost allocated to a Product} = (\text{Product output proportion}) \times (\text{Total joint cost})$$

$$\text{Joint cost allocated to Product X} = 0.375 \times \$16,00 = \$6,000$$

$$\text{Joint cost allocated to Product Y} = 0.625 \times \$16,00 = \$10,000$$

If it is desired, the average unit cost for each product can be determined by dividing its allocated cost by its production quantity, as follows:

$$\text{Average unit cost for Product X} = (\$6{,}000)/(1{,}200 \text{ gallons}) = \$5.00 \text{ per gallon}$$

$$\text{Average unit cost for Product Y} = (\$10{,}000)/(2{,}000 \text{ gallons}) = \$5.00 \text{ per gallon}$$

As you have noticed from this example, the joint costs per unit for all products are the same. However, the final manufacturing will be different as the separable costs (per unit) are different among the products.

The following is a partial product-line income statement for the above example.

AAA Chemical Company
Product-Line Income Statement
(Physical-Measure Method for Allocating Joint Costs)

	Product X	Product Y	Total
Sales	(900 × \$15) = \$13,500	(1,500 × \$3) = \$4,500	\$18,000
Joint Costs	(900 × \$5) = 4,500	(1,500 × \$5) = 7,500	12,000
Gross Margin	\$9,000	(\$3,000)	\$6,000
Gross Margin as a % of Sales	66.67	−66.67	33.33

This example unveiled a major disadvantage of the physical-measure method of cost allocation, which is an unfair cost assignment to Product Y whose market value is not even close to that of Product X. The method is practical only for the production processes in which the difference among the market values of the joint products is very insignificant. However, if a process produces joint products that substantially differ in values from each other, the physical-measure method would not be appropriate for the joint costs allocation.

5.5.2. Weighted-Average Method of Joint Cost Allocation

In some situations, the simple-average method may not give a satisfactory result for the joint cost apportionment. The **weighted-average method** would often provide a better solution to the joint cost allocation problem. The weighted-average method uses weight factors assigned to each unit of the joint products, such as the size of the product, its complexity and difficulty to manufacture, time consumed in manufacturing the unit, the type of labor skills, amount of materials used, etc.

Under the weighted-average method, the finished production quantity of each product is multiplied by a weight factor as a basis for allocating the joint cost to the individual products.

Example 5.3

In Example 5.2, assume that AAA Chemical Company uses the weighted-average method for allocating its joint cost to Products X and Y with the following weight factors:

Product X: 4.5 points
Product Y: 1.3 points

The total joint cost (JC) to be allocated to the joint products is \$16,000.

Solution

Weighted units:

$$\text{For Product X} = 4.5(1{,}200 \text{ gallons}) = 5{,}400 \text{ gallons}$$

$$\text{For Product Y} = 1.3(2{,}000 \text{ gallons}) = 2{,}600 \text{ gallons}$$

Joint cost per weighted unit:

$$\text{Total weighted units} = 5{,}400 + 2{,}600 = 8{,}000 \text{ gallons}$$

$$\$16{,}000/8{,}000 = \$2.00 \text{ per weighted unit}$$

Allocated joint cost:

$$\text{To Product X} = \$2(5{,}400) = \$10{,}800$$

$$\text{To Product Y} = \$2(2{,}600) = \$5{,}200$$

Joint cost per unit:

$$\text{For Product X} = \$10{,}800/1{,}200 = \$9.00 \text{ per gallon}$$

$$\text{For Product Y} = \$5{,}200/2{,}000 = \$2.60 \text{ per gallon}$$

The following is a partial product-line income statement in this example.

AAA Chemical Company
Product-Line Income Statement
(Physical-Measure Method for Allocating Joint Costs)

	Product X	Product Y	Total
Sales	900 × \$15 = \$13,500	1,500 × \$3 = \$4,500	\$18,000
Joint Costs	900 × \$9 = 8,100	1,500 × \$2.60 = 3,900	12,000
Gross Margin	\$5,400	\$600	\$6,000
Gross Margin as a % of Sales	40.00	13.33	33.33

As you noticed, Product Y was not profitable under the simple-average-cost (i.e., the physical-measure) method, but under this method, it becomes profitable because less joint cost is allocated to it.

5.5.3. Sales-Value Method of Joint Cost Allocation

The sales-value method is used when the joint products are sold at the split-off point "as is" without further processing (i.e., there are no separable costs). This method is similar to the weighted-average method. It uses the sales values of the products as the basis (as a weight factor) for joint cost allocation. Under the sales-value method, the user allocates the joint cost in the following steps:

Step 1: Compute the ***potential*** sales value (SV_i) for each of the joint products:

$$\mathbf{SV}_i = (\textbf{Sales Price}) \times (\textbf{Production Quantity})$$

where $i = 1, 2, \ldots, n$ is product index.

Step 2: Compute the **total sales value** (**TSV**) by summing the sales values of all joint products, from **Step 1**.

$$\mathbf{TSV} = \sum_{i=1}^{n} \mathbf{SV}_i.$$

Step 3: Compute the **proportion** of sales value of total sales value (TSV) for each product:

$$\textbf{Proportion for product } i = \mathbf{SV}_i / \mathbf{TSV}.$$

Step 4: Allocate the joint cost to each joint product by multiplying its corresponding proportion by the joint cost:

$$\textbf{Joint cost allocated to product } i = (\mathbf{SV}_i / \mathbf{TSV}) \times (\textbf{Joint cost}).$$

Example 5.4

In Examples 5.2 and 5.3, assume that AAA Chemical Company uses the sales-value method for allocating its joint cost to Products X and Y and sells the product without any further processing at \$15 per gallon of Product X and \$3 per gallon of Product Y.

Solution

Step 1: Compute the *potential* sales value (SV_i) for each of the joint products:

$$SV_X = \$15 \times 1{,}200 \text{ gallons} = \$18{,}000$$

$$SV_Y = \$3 \times 2{,}000 \text{ gallons} = \$6{,}000$$

Step 2: Compute the total sales value (TSV) by summing the sales values of all joint products, from **Step 1**.

$$TSV = \$18,000 + \$6,000 = \$24,000$$

Step 3: Compute the proportion of sales value of total sales value (TSV) for each product:

$$\text{Proportion for Product X} = \$18,000/\$24,000 = 0.75 \text{ or } 75\%$$

$$\text{Proportion for Product Y} = \$6,000/\$24,000 = 0.25 \text{ or } 25\%$$

Step 4: Allocate the joint cost to each joint product by multiplying its corresponding proportion by the joint cost:

$$\text{Joint cost allocated to Product X} = 0.75(\$16,000) = \$12,000$$

$$\text{Joint cost allocated to Product Y} = 0.25(\$16,000) = \$4,000$$

Joint cost per unit:

$$\text{For Product X} = \$12,000/1,200 = \$10.00 \text{ per gallon}$$

$$\text{For Product Y} = \$4,000/2,000 = \$2.00 \text{ per gallon}$$

The following is a partial product-line income statement in this example.

AAA Chemical Company
Product-Line Income Statement
(Physical-Measure Method for Allocating Joint Costs)

	Product X	Product Y	Total
Sales	900 × $15 = $13,500	1,500 × $3 = $4,500	$18,000
Joint Costs	900 × $10 = 9,000	1,500 × $2.00 = 3,000	12,000
Gross Margin	$4,500	$1,500	$6,000
Gross Margin as a % of Sales	33.33	33.33	33.33

As you notice, both products would generate equal gross margin percentage under this method since the allocated joint cost to each product is in proportion to their sales value.

5.5.4. Relative-Market-Value Method of Joint Cost Allocation

At the split-off point, although the products are separated, they may not be salable because they are not finished yet. Thus, finishing the products require further (separate) manufacturing or processing, which incur further processing (or separable) costs. There may also be a separable selling costs for each product in addition to the separable manufacturing costs.

Figure 5.1, as presented at the beginning of this chapter, illustrates the relationship among joint costs, separable costs, and joint products.

The relative-market-value method (also known as the net-realizable-value method) allocates joint costs to products in proportion to their relative sales values at the split-off point. The relative market (or net realizable) values means approximated sales values of the joint products at the split-off point, which is computed by subtracting the further processing costs (incurred beyond the split-off point) from the estimated final sales values. If no additional processing is required beyond the split-off point, the method becomes the sales-value method (i.e., the previous method).

If the physical-measure or weighted-average method of joint cost allocation is used, there is no need to know the relative market values of the products at the split-off point. However, if we should use the relative-market-value method to allocate joint costs, we will have to approximate the market value of each joint product at the split-off point.

Under the sales-value method, the user allocates the joint cost in the following steps:

Step 1: Compute the net realizable (relative market) value (RMV_i or NRV_i) for each of the joint products:

$$\mathbf{NRV} = \left(\mathbf{Sales\ Price} - \mathbf{Separable\ \ cost\ per\ unit}\right) \times \left(\mathbf{Production\ Quantity}\right)$$

$$\mathbf{NRV}_i = \left(\boldsymbol{P_i} - \mathbf{SC}\boldsymbol{i}\right) \times \boldsymbol{Q_i}$$

where $i = 1, 2, \ldots, n$ is product index.

Step 2: Compute the **total net realizable value** (**TNRV**) by summing the net realizable values computed in **Step 1**.

$$\mathbf{TNRV} = \sum_{i=1}^{n} \mathbf{NRV}_i.$$

Step 3: Compute the **proportion** of the net realizable value for each product relative to the total net realizable value (TNRV):

$$\mathbf{Proportion\ for\ product}\ \boldsymbol{i} = \mathbf{NRV}_i\ /\ \mathbf{TNRV}.$$

Step 4: Allocate the joint cost to each joint product by multiplying its corresponding proportion by the joint cost:

$$\textbf{Joint cost allocated to product } i = \left(\textbf{NRV}_i \,/\, \textbf{TNRV}\right) \times \left(\textbf{Joint cost}\right).$$

Example 5.5

The XYZ Chemical Company produces two products, A and B, using a single input material in its manufacturing process.

Assumptions

- No market exists for either product at the split-off point.
- Additional manufacturing and selling costs must be incurred prior to the final point, where both products become ready for sale.
- During the month of June, 50,000 gallons of materials were processed at a joint cost of $100,000.
- All units have been completed at the end of June.
- The following table shows how many gallons of each product resulted from the process, and for each product, the quantity sold, the selling price per unit, as well as the separable cost per unit.

	Product A	Product B
Production (in gallons)	25,000	25,000
Sales (in gallons)	20,000	15,000
Ending Inventory – gallons	5,000	10,000
Sales Price – per gallon	$8.00	$14.00
Separable Costs:		
Manufacturing – per gallon	$3.00	$6.00
Selling – per gallon	$1.00	$2.00

The provided data can be visualized by the following diagram:

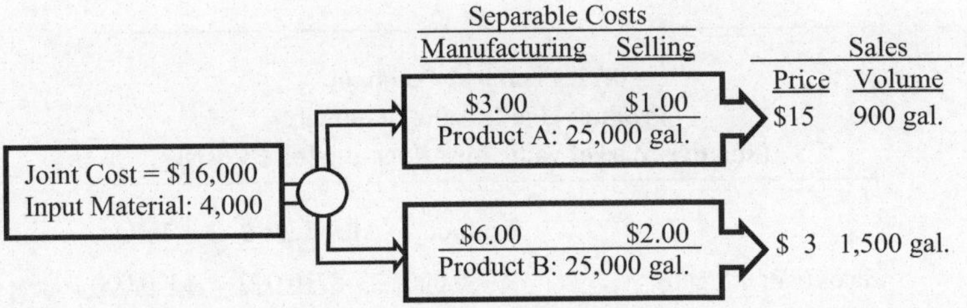

Since no actual sales price exists at the split-off point for Products A and B, we must first approximate the relative market values of the total production volume for each of the joint products at the split-off point. Then, we can allocate the joint cost.

	Product A	Product B	Total
Product Volume (gallons)	25,000	25,000	50,000
Sales Price – per gallon	$8.00	$14.00	—
Total Sale Value of Products	$200,000	$350,000	$550,000
Less: Separable costs:			
Manufacturing	(75,000)	(150,000)	(225,000)
Selling	(25,000)	(50,000)	(75,000)
Approximate Market Value at Split-Off Point	$100,000	$150,000	$250,000
Proportion	100/250 = 40%	150/250 = 60%	100%
Share of joint cost	**$40,000**	**$60,000**	**$100,000**

To obtain the total manufacturing cost for the purpose of calculating the manufacturing unit cost for each product, add the separable manufacturing cost (but not the separable selling expenses) to the allocated joint cost for each product, as shown below:

	Product A	Product B	Total
Allocated Joint Costs	$40,000	$60,000	$100,000
Add: Separable Manufacturing costs	75,000	150,000	225,000
Total Manufacturing Costs	$115,000	$210,000	$325,000
Production Volume, gallons	25,000	25,000	—
Unit Cost (cost per gallon)	$4.60	$8.40	—

The separable selling costs (expenses) are not included since they are periodic costs that are expensed in the period they are expired. The costs per unit, $4.60/gallon and $8.40/gallon, are used to value the ending inventories of Products A and B, respectively.

XYZ Chemical Company
Product-Line Income Statement
(Relative Market Value for Allocating Joint Costs)

	Product A	Product B	Total
Sales (gal. × $/gal.)	$160,000	$210,000	$370,000
Manufacturing Costs*	$115,000	$210,000	$325,000
Less: Ending Inventory**	23,000	84,000	107,000
Cost of Goods Sold	$92,000	$126,000	$218,000
Gross Margin	**$68,000**	**$84,000**	**$152,000**
Gross Margin as % of Sales	42.5	40	41.08
Less: Period Expenses***	$20,000	$30,000	$50,000

Profit before Taxes	**$48,000**	**$54,000**	**$102,000**
Profit per Unit	$2.40	$3.60	—

* Total manufacturing costs for all units.
** Ending Inventory: A: 5,000 gallons×$4.60/gal. = $23,000
B: 10,000 gallons×$8.40/gal. = $84,000
*** Selling Expense: A: ($25,000/25,000 gal.)×20,000 gals. sold = $20,000
B: ($50,000/25,000 gal.)×15,000 gals. sold = $30,000

5.5.5. Constant-Gross-Margin Method of Joint Cost Allocation

This method allocates joint costs to joint (main) products in such a way that each individual product attains the same gross-margin ratio, rate, or percentage (GMR or GM%). This ratio is computed as follows:

$$\text{Gross Margin} = \text{Sales Value} - \text{Cost of Goods Sold}$$

$$\text{GM\%} = \frac{\text{Gross Margin}}{\text{Sales}}$$

Note that, in joint cost allocation, the sales value is the potential sales values of all joint products produced during a period (not just the units of products that were actually sold during the period), and the cost of goods sold would be the total of joint costs and all separable costs.

Under the constant-gross-margin method, the user allocates the total joint cost in 5 steps:

Step 1: Compute the potential sales value (ultimate sales value) for all and each product, which the products would eventually earn.

Step 2: Compute the total cost by summing the joint costs and all separable costs for all joint products, as expressed by the following formula:

$$\textbf{Total Cost} = \textbf{Joint Costs} + \textbf{Total Separable Costs}$$

Step 3: Compute the overall gross margin by subtracting the total costs from the total potential revenues, as expressed by the following formula:

$$\textbf{Overall Gross Margin} = \textbf{Total Final Sales Value of All Products} - \textbf{Total Costs}$$

Step 4: Compute the overall gross margin percentage by dividing the total gross margin by the total potential revenues, as expressed by the following formula:

$$\textbf{Gross Margin Percentage (or Ratio)} = \frac{\textbf{Overall Gross Margin}}{\textbf{Total Sales Value of All Products}}$$

Step 5: Allocate joint cost to each product by multiplying the gross margin percentage by the potentials sales value of the product and then deducting its separable cost, as expressed by the following formula:

$$\textbf{Joint Costs Allocated to a Product} = (\textbf{Sales Value of the Product}) - (\textbf{Gross Margin of the Product}) - (\textbf{Separable Cost of the Product})$$

Example 5.6

During the previous month, the XYZ Company produced three products through a single process and incurred a joint production cost of $350,000. The following relevant financial data for the month are also provided:

Product	Production Level	Separable Cost	Sales Price
X	12,500	$120,000	$32
Y	6,500	100,000	50
Z	7,000	60,000	25

Required: Allocate the joint production cost to Products X, Y, and Z using the constant-gross-margin-percentage method.

Solution

Step 1: Compute the total potential revenue (TR):

$$TR = \$32(12{,}500) + \$50(6{,}500) + \$25(7{,}000)$$

$$= \$400{,}000 + \$325{,}000 + \$175{,}000 = \$900{,}000$$

Step 2: Compute the total cost (TC):

$$TC = \$350{,}000 + (\$120{,}000 + \$100{,}000 + \$60{,}000) = \$630{,}000$$

Step 3: Compute overall gross margin (GM)

$$GM = \$900{,}000 - \$630{,}000 = \$270{,}000$$

Step 4: Compute the gross margin percentage (GM%)

$$GM\% = \$270{,}000 \,/\, \$900{,}000 = 0.3 = 30\% \text{ of revenue}$$

Step 5: Allocate joint cost to the joint products using GM%

(1)	(2)	(3)	(4) = (2)×(3)	(5) = (2)−(4)	(6)	(7) = (5)−(6)
Product	**Potential Revenue**	**GM%**	**Gross Margin**	**Total Cost**	**Separable Cost**	**Allocated JC**
X	$400,000	30	$120,000	$280,000	$120,000	$160,000
Y	325,000	30	97,500	227,500	100,000	127,500
Z	175,000	30	52,500	122,500	60,000	62,500
Total	$900,000	30	$270,000	$630,000	$280,000	$350,000

The manufacturing unit cost (UC) for the products are:

UC_X = $280,000/12,500 = $22.40
UC_Y = $227,500/6,500 = $35.00
UC_Z = $122,500/7,000 = $17.50

5.6. TRADITIONAL METHODS FOR JOINT COST ALLOCATION TO BY-PRODUCTS

Because the market value of by-products is not significant as compared to the market values of the main products, accountants usually follow simple methods in accounting for by-products. The two commonly used methods by accountants are (1) net-realizable-value (or production) method and (2) zero-cost (or sales) method.

5.6.1. Net-Realizable-Value or Production Method

One method of accounting for by-products is the net-realizable-value (NRV) or production method. A firm may use this method to value the quality of by-products produced at their net realizable value at the split-off point. Net realizable value is determined by the final sales price less separable costs. This method has the following features:

- The joint cost assigned to a by-product equals its net realizable value (NRV) at the split-off point. The joint cost assigned to the by-product reduces the amount of joint costs assigned to the main products (i.e., the NRV of the by-product is treated as cost savings).
- The by-product is recognized in the period in which it is produced.
- The cost of then by-product is added to the "By-product Inventory" account.
- When the by-product is sold, as its sales value and cost are equal to its NRV, it will not generate any profit.

Accountants debit (charge) the total net realizable value of the by-products into the "By-product Inventory" account and credit the "Joint Costs" account. Therefore, joint costs of producing the main products are reduced by the net realizable value of the by-products. The corresponding journal entry is usually recorded as follows:

By-product Inventory	xxx	
Joint Costs		xxx

Example 5.5

Suppose that, during a period, a company by processing 13,700 kg of a certain material jointly produced 9,000 kg of Product A, 4,000 kg of Product B, and 500 kg of a by-product (with some waste material). Assume that the company can sell the product, at the split-off point without any further processing, as Product A at \$4.00/kg, Product B at \$3.00/kg, and the by-product at \$0.40/kg. The situation can be diagrammed as:

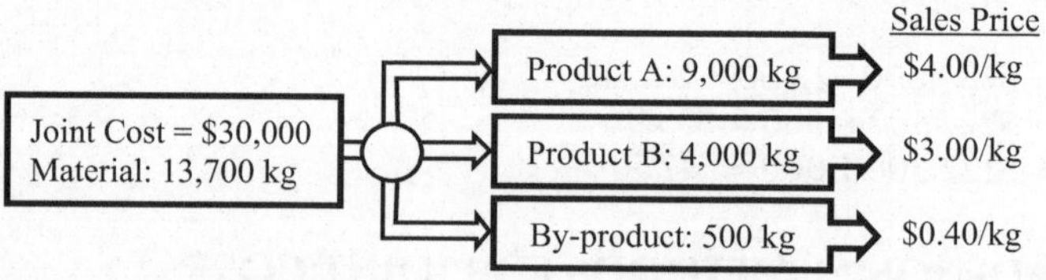

Assume, also, that company incurred the following joint costs:

Material costs	\$12,000
Labor costs	\$10,000
Predetermined overhead rate is 80% of the labor costs, that is, 10,000 × 80%	\$8,000
Total joint costs	**\$30,000**

In this example, the net realizable value of the by-product (NRV_B) is equal to the sales price since no additional costs are necessary to complete and sell the by-product. Then,

NRV_B = \$0.40(500) = **\$200** (joint cost assigned to the by-product)
Joint cost to be assigned to main products = \$30,000 − \$200 = **\$29,800**

The journal entries to record the transactions will be as follows:

Joint Cost	30,000	
Materials Inventory		12,000
Wage Payable		10,000
Applied Overhead		8,000
To record the incurred joint costs		
By-product Inventory	200	
Joint Cost		200
To record the assigned cost to the by-product		

If the firm uses the sales-value method, it would allocate the joint costs to the main products as summarized in the following cost allocation schedule:

Product	Production Level (kg)	Selling Price at Split-off Point	Market Value at Split-off Point	Proportion*	Joint Cost to be Allocated	Share of Joint Costs
A	9,000	$4.00	$36,000	75%		$22,350
					$29,800	
B	4,000	$3.00	$12,000	25%		$ 7,450
Total			**$48,000**	**100%**	**$29,800**	**$29,800**

* 36,000/48,000 = 75% and 12,000/48,000 = 25%.

The unit cost for Products A and B and the by-product (BP) would be:

Product	Computation	Cost per kg
A	$22,350/9,000	$2.4833
B	$7,450/4,000	$1.8625
BP		$0.4000

5.6.2. Zero-Cost Method

The second method for accounting for by-products is the zero-cost or sale method. This method assumes that no joint cost will be allocated to a by-product, so named as the zero-cost method. Therefore, the entire joint cost, in the previous example $30,000 (instead of $29,800), would be allocated to the two main products. Thus, when a by-product is sold, its sales amount would be added to "revenues" in the income statement and the cost of the by-product sold would be zero if there is no further (separable) processing cost.

If separable processing costs are incurred to complete a by-product, the first method would add the separable costs incurred for the by-product to its net realizable value.

If a firm uses the zero-cost method and does not assign any joint cost to the by-product at the split-off point then only the separable costs would be debited to the "By-product Inventory" account and credited to the corresponding accounts (i.e., materials and labor costs and applied overheads incurred for further processing), as shown below:

By-product Inventory	xxx	
Materials Inventory		xxx
Direct Labor		xxx
Applied Overhead		xxx
To record by-product separable costs		

The sales of the by-product are either added to the sales of the other joint products or deducted from the cost of goods sold (COGS) of the other joint products. The sales or COGS adjustments are done on the financial statement and not in the accounting records.

5.7. QUANTITATIVE METHOD FOR JOINT COST ALLOCATION

The disagreement among practitioners in using joint cost allocation techniques has made their techniques arbitrary and sometimes unjustifiable (Manes and Smith, 1965). To overcome this issue, engineers and quantitative analysts have applied mathematical programming to joint cost allocation and by-product identification (e.g., Manes and Smith, 1965; Weil, 1968; Kaplan, 1982; and Tayyari and Parsaei, 1992). Engineering (mathematical programming) techniques are useful in giving an insight into the nature of the joint-product problems and production optimization. They also provide a basis for distinction between joint-product and by-product.

5.7.1 Formulating Joint Cost Allocation

Tayyari and Parsaei (1992) presented a nonlinear programming model for joint cost allocation to joint products as follows.

Let N be the set of n sequenced numbers representing the product indexes, and X_i ($i = 1, 2, \ldots, n$) be the sales quantity of the product i. The following assumptions are made for the purposes of formulating the joint cost allocation:

1) $\boldsymbol{n}$ products are simultaneously produced in fixed proportions. That is, processing each unit of the input material yields a_i units of the $\boldsymbol{i}$th product.
2) The production process is subject to a limited amount of $\boldsymbol{K}$ units of the input material.
3) There exists enough information about the demand functions for the joint products.

Hence, there will be n constraints representing the relationships between sales and the input material quantities (i.e., $X_i \leq a_i Y$, when Y units of input material are processed and a_i units of product i will be produced, where $i = 1, 2, \ldots, n$).

Letting P_i indicate the price of the ith product ($i = 1, 2, \ldots, n$) and X_i the number of units of this product produced and sold, P_i can be expressed as a function of the demand quantity. This function can be denoted as:

$$P_i = g_i\left(X_i\right), \quad i = 1,2,\ldots,n \tag{5.1}$$

Let's assume that Y units of the input (raw) material are processed at a cost of C dollars per unit (material and other processing variable costs). If X_i units of product i ($i = 1, \ldots, n$) are sold, the total revenue, TR($\mathbf{X}$), and total cost, TC($\mathbf{X}$), are written as:

$$\mathrm{TR}(\mathbf{X}) = \sum_{i=1}^{n} X_i P_i + \sum_{i=1}^{n} X_i g_i(X_i) \tag{5.2}$$

$$\mathrm{TC}(\mathbf{X}) = \mathrm{TC}(Y) = C \cdot Y$$

(**Note**: $\mathbf{X} = (X_1, X_1, \ldots, X_n)$, is the sales vector.)

The contribution margin is calculated as:

$$Z(\mathbf{X}, Y) = \mathrm{TR}(\mathbf{X}) - \mathrm{TC}(Y) \tag{5.3}$$

Therefore, the decision problem is to maximize the contribution margin (which guarantees the maximum profits) subject to the availability and non-negativity constraints. That is,

$$\text{Maximize: } Z(\mathbf{X}, Y) = \sum_{i=1}^{n} X_i g_i(X_i) - C \cdot Y \tag{5.4}$$

$$\begin{aligned} \text{Subject to: } & X_i \le a_i Y,\ i = 1, \ldots, n && (\text{sales constraints}) \\ & Y \le K, && (\text{material constraint}) \\ & X_i, Y \ge 0,\ i = 1, \ldots, n && (\text{non-negativity constraints}) \end{aligned}$$

This is a nonlinear programming problem that may be solved by classical quantitative techniques in operations research. For example, the Lagrangian methods can be used to obtain an optimal solution. This method will be particularly simple when the demand–price functions are assumed to be linear. In such a case, the objective function would only involve linear and squared variables, and the problem, then, called a quadratic programming problem.

Kaplan (1982) has applied the Lagrangian method in some numerical examples with linear demand functions and relaxed the material constraint. By doing so, which seems to be practical, and assuming $P_i = b_i - m_i X_i$ (both b_i and m_i are non-negative constant coefficients of the price–demand function for product i, where $i = 1, 2, \ldots, n$), the above nonlinear programming problem is reduced to the following form:

$$\text{Maximize: } Z(\boldsymbol{X}, Y) = \sum_{i=1}^{n} X_i (b_i - m_i X_i) - C \cdot Y \tag{5.5}$$

$$\begin{aligned} \text{Subject to: } & a_i Y - X_i \ge 0,\ i = 1, \ldots, n && (\text{sales constraints}) \\ & X_i, Y \ge 0,\ i = 1, \ldots, n && (\text{non-negativity constraints}) \end{aligned}$$

Under the Lagrangian method, each constraint is assigned a non-negative Lagrangian multiplier, λ_i, such that $\lambda_i(a_iY - X_i) = 0$, and add all $\lambda_i(a_iY - X_i)$ to the objective function to obtain the following form of nonlinear programming with non-negativity constraints only:

$$\text{Maximize: } Z(\boldsymbol{X},Y,\lambda) = \sum_{i=1}^{n} X_i\left(b_i - m_iX_i\right) - C\cdot Y + \sum_{i=1}^{n}\left(a_iY - X_i\right)_i \quad (5.6)$$

$$\text{Subject to: } X_i, Y, \lambda_i \geq 0, i = 1,\ldots,n \qquad (\text{non-negativity constraints})$$

Now, the optimal solution can be obtained by using Kuhn-Tucker conditions or simply setting the first partial derivative of L(**X**, *Y*, **λ**) with respect to each variable (i.e., X_i, Y, λ_i) equal to zero and then solving the following resulting equations simultaneously:

$$\frac{\partial L}{\partial X_i} = b_i - 2m_iX_i - \lambda_i = 0 \quad i \in N \quad (5.7)$$

$$\frac{\partial L}{\partial Y} = -C + \sum_{i=1}^{N} a_i\lambda_i = 0 \quad i \in N \quad (5.8)$$

$$\frac{\partial L}{\partial_i} = a_iY - X_i = 0 \quad i \in N \quad (5.9)$$

$$X_i, Y, \lambda_i \geq 0 \quad (5.10)$$

Given that all conditions expressed by Equations 5.7 through 5.10 are satisfied, the results will be the optimal solution to the problem and appear in the following forms:

$$Y = \frac{-C + \sum_{i}^{n} a_ib_i}{2\sum_{i=1}^{n} m_ia_i^2} \quad (5.11)$$

Optimal number of units of input material to be processed

$$X_j = a_j\left[\frac{-C + \sum_{i}^{n} a_ib_i}{2\sum_{i=1}^{n} m_ia_i^2}\right] \quad \text{Production and sales units of product } j \quad (5.12)$$

$$\lambda_j = b_j + m_ja_j\left[\frac{C - \sum_{i}^{n} a_ib_i}{\sum_{i=1}^{n} a_i^2}\right] \quad \text{Joint cost assigned to each unit of product } j \quad (5.13)$$

In Equation 5.8, $\left[\sum_{i=1}^{n} a_i\lambda_i = C\right]$, C is the joint cost allocated to the n products of which λ_i dollars are allocated to each unit of product i.

However, if any of the λ_i is found to be negative, which violates the non-negativity conditions, it must arbitrarily be set equal to zero and the corresponding product be treated as a by-product, on a zero-cost basis. Then, the value $\lambda_i = 0$ is substituted in Equations 5.7 and 5.8, the corresponding equation in the Equation set 5.9 is eliminated, and the revised equations are solved again. In such circumstances, some units of product i will remain unsold, which should be discarded. Otherwise, selling all units of that by-product will reduce the price as well as the overall profits, unless the residuals are sold in a different market.

Example 5.6

As an example, consider a situation in which for each unit of an input raw material, one unit of Product 1 and two units of Product 2 are obtained. Assume that the price of each product is affected by the volume of product supplied to the market. Let

P_i = the price of product i, where $i = 1, 2$, and

Q_i = the volume of product i produced and sold.

Assume the following relations for product's volume and price:

$$P_1 = 15 - Q_1 \tag{5.14}$$

$$P_2 = 49 - 2Q_2 \tag{5.15}$$

Assume, also, that the input material is bought at a price of \$6/unit. Variable cost is \$4/unit of processed input material. The question is how many units of the input material should be bought and processed.

Let Y be the number of units of the input material to be processed. Then, Y units of Product 1 and $2Y$ units of Product 2 can be produced. The total revenue will be

$$\text{TR} = P_1Q_1 + P_2Q_2$$

Knowing that P_1 is a function of Q_1 and P_2 is a function of Q_2, substituting P_1 and P_2 by their equivalent values in terms of Q_1 and Q_2 from Equation (5.14) and (5.15), respectively, the total revenue will be a function of Q_1 and Q_2.

$$\text{TR} = Q_1 \cdot (15 - Q_1) + Q_2 \cdot (49 - 2Q_2)$$

or

$$\text{TR} = 15Q_1 - Q_1^2 + 49Q_2 - 2Q_2^2$$

Therefore, the objective will be to maximize the contribution margin (total revenue minus processing variable cost):

$$\text{Maximize: } Z = 15Q_1 - Q_1^2 + 49Q_2 - 2Q_2^2 - (6+4)Y$$

or

$$\text{Maximize: } Z = 15Q_1 - Q_1^2 + 49Q_2 - 2Q_2^2 - 10Y$$

$$\text{subject to: } Q_1 \le Y$$
$$Q_2 \le 2Y$$
$$Q_1, Q_2, Y \ge 0$$

This is a nonlinear programming with two inequality constraints and non-negativity condition.

This problem can be solved using Lagrange multipliers. That is, for each inequality constraint, a non-negative Lagrange multiplier, λ_i, will be associated:

λ_1 for $Q_1 \le Y$

λ_2 for $Q_2 \le 2Y$

If $Q_1 = Y$, then $\lambda_1 > 0$ and if $Q_1 < Y$, then $\lambda_1 = 0$.

If $Q_2 = 2Y$, then $\lambda_2 > 0$ and if $Q_2 < 2Y$, then $\lambda_2 = 0$.

Here, the Lagrange multipliers play roles of dual variables (as for linear programming). The problem would be:

$$\text{Maximize: } L = 15Q_1 - Q_1^2 + 49Q_2 - 2Q_2^2 - 10Y + \lambda_1(Y - Q_1) + \lambda_2(2Y - Q_2)$$

$$\text{subject to: } Q_1, Q_2, Y, \lambda_1, \lambda_2 \ge 0.$$

To solve this problem, the first partial derivatives will be set equal to zero (the first-order conditions):

$$\frac{\partial L}{\partial Q_1} = 15 - 2Q_1 - \lambda_1 = 0 \tag{5.16}$$

$$\frac{\partial L}{\partial Q_2} = 49 - 4Q_2 - \lambda_2 = 0 \tag{5.17}$$

$$\frac{\partial L}{\partial Y} = -10 + \lambda_1 + 2\lambda_2 = 0 \tag{5.18}$$

$$\frac{\partial L}{\partial_1} = Y - Q_1 = 0 \tag{5.19}$$

$$\frac{\partial L}{\partial_2} = 2Y - Q_2 = 0 \tag{5.20}$$

$$Q_1 = Y$$

$$Q_2 = 2Y$$

$$15 - 2Y - \lambda_1 = 0 \quad \Rightarrow \lambda_1 = 15 - 2Y$$

$$49 - 4(2Y) - \lambda_2 = 0 \quad \Rightarrow \lambda_2 = 49 - 8Y$$

$$-10 + \lambda_1 + 2\lambda_2 = 0$$

$$-10 + (15 - 2Y) + 2(49 - 8Y) = 0 \quad \text{or}$$

$$18Y = 103 \quad \Rightarrow Y = 5.72 \text{ units}$$

$$Q_1 = Y \quad \Rightarrow Q_1 = 5.72 \text{ units}$$

$$Q_2 = 2Y = 2(5.72) \quad \Rightarrow Q_2 = 11.44 \text{ units}$$

$$\lambda_1 = 15 - 2Y = 15 - 2(5.72) \quad \Rightarrow \lambda_1 = \$3.56$$

$$\lambda_2 = 49 - 8Y = 49 - 8(5.72) \quad \Rightarrow \lambda_2 = \$3.24$$

$$P_1 = 15 - X_1 \quad \Rightarrow P_1 = \$9.28$$

$$P_2 = 49 - 2X_2 \quad \Rightarrow P_2 = \$37.56$$

Equation (5.18) can be rewritten as:

$$\lambda_1 + 2\lambda_2 = 10 \tag{5.21}$$

where 10 is the joint cost to be allocated to Products A and B. Equation (5.21) provides the following cost allocation to the products:

λ_1 dollars for each unit of product 1, and
λ_2 dollars for each unit of product 2.

It is known, from economic theory and elementary calculus, that achieving the maximum profit requires that the marginal revenue be equal to the marginal cost. The marginal revenues of the two products in this example are:

$$MR_1 = \frac{\partial}{\partial Q_1}(P_1 Q_1) = 15 - 2Q_1$$

$$MR_2 = \frac{\partial L}{\partial Q_2}(P_2Q_2) = 49 - 4Q_2$$

The numerical values of marginal revenues at the optimum values (i.e., $Q_1 = 5.72$ units and $Q_2 = 11.44$ units) are:

$$MR_1 = 15 - 2(5.72) = \$3.56$$

$$MR_2 = 49 - 4(11.44) = \$3.24$$

The marginal cost of processing one unit of the input material (which produces one unit of Product 1 and two units of Product 2) is \$10. The marginal revenue obtained through processing an additional unit of the input material is:

$$MR_1 + 2MR_2 = 3.56 + 2(3.24) = \$10.04$$

This is approximately \$10, which is equal to the marginal cost. This result verifies that the computations for the optimal solution were correct.

5.7.2. A Basis for Identifying By-products

Consider a situation in which the processing of 1 unit of raw material yields 3 units of Product 1 and 2 units of Product 2. The demands for the two products in the market as functions of sales prices are given as follows:

$$Q_1 = 36 - P_1 \qquad \text{(for Product 1)}$$

$$Q_2 = 19 - 0.5P_2 \qquad \text{(for Product 2)}$$

Assume that the supply of the raw material is unlimited and it costs \$8 per unit. The variable cost of processing one unit of the raw material is \$6 per unit. Find the optimal quantity of the raw material and quantities and prices of the output products. The problem is a nonlinear program.

$$P_1 = 36 - Q_1 \Rightarrow P_1Q_1 = 36Q_1 - Q_1^2$$

$$P_2 = 38 - 2Q_2 \Rightarrow P_2Q_2 = 38Q_2 - 2Q_2^2$$

Then the Lagrangian model is of the form:

$$\text{Max } L = 36Q_1 - Q_1^2 + 38Q_2 - 2Q_2^2 - 14Y + \lambda_1(3Y - Q_1) + \lambda_2(2Y - Q_2)$$

$$\text{Subject to: } Q_1, Q_2, Y, \lambda_1, \lambda_2 \geq 0$$

The first-order conditions of the problem are:

$$\frac{\partial L}{\partial Q_1} = 36 - 2Q_1 - \lambda_1 = 0$$

$$\frac{\partial L}{\partial Q_2} = 38 - 4Q_2 - \lambda_2 = 0$$

$$\frac{\partial L}{\partial Y} = -14 + 3\lambda_1 + 2\lambda_2 = 0$$

$$\frac{\partial L}{\partial_1} = 3Y - Q_1 = 0$$

$$\frac{\partial L}{\partial_2} = 2Y - Q_2 = 0$$

From the last two equations $Q_1 = 3Y$ and $Q_2 = 2Y$. By substituting these values for Q_1 and Q_2 in the first two equations and solving for λ_1 and λ_2 in terms of Y, the following would be the result: $\lambda_1 = 36 - 6Y$ and $\lambda_2 = 38 - 8Y$.

Now, substitute these values for λ_1 and λ_2 in the third equation:

$$-14 + 3(36 - 6Y) + 2(38 - 8Y) = 0 \Rightarrow Y = 5$$

The other factors are:

$$Q_1 = 15, Q_2 = 10, \lambda_1 = 6, \lambda_2 = -2 < 0.$$

The value $\lambda_2 = -2 < 0$ violates the non-negativity constraint of the problem. Therefore, the solution is not acceptable. This shows that the manufacturer should sell less quantity of Product 2 than what he produces. Also the marginal revenue of product 2 (i.e., $MR_2 = 38 - 4Q_2$) will be zero at $Q_2 = 9.5$, that is, the manufacturer should not sell more than 9.5 units of this product.

Since $Q_2 = 9.5 < 10 = 2Y$, then $\lambda_2 = 0$ and another solution to the first-order conditions must be found (given these values).

$$36 - 2Q_1 - \lambda_1 = 0$$

$$-14 + 3\lambda_1 = 0$$

$$3Y - Q_1 = 0$$

Thus the optimal solution is:

$$Q_1 = 47/3 = 15.6 \text{ units}, Q_2 = 9.5 \text{ units}, Y = 47/9 = 5.2 \text{ units},$$

$$\lambda_1 = \$14/3 = \$4.67, \lambda_2 = \$0, P_1 = \$20.40, \text{ and } P_2 = \$19.00$$

Noting that 0.9 unit of Product 2 from the total of 2(5.2) = 10.4 units produced should be discarded.

Since some of the Product 2 is discarded, it must be a by-product. The dual variable, λ_2, associated with Product 2 is also zero, that is, no joint cost is allocated to it.

Hence, by-products are those products whose associated nonlinear programming (Lagrangian) dual variables are zero.

SUMMARY

Joint costs are allocated to products that are produced through a common process or from the same input material. The main objective of joint cost allocation is to fairly relate the costs of the inputs to the joint products. As there is no direct method for such cost allocation, some approximations become inevitably necessary.

The following are several methods that are commonly used for allocating joint cost (or common cost) to products that are produced jointly or through the same process. Where joint products are sold at the split-off point, one of the following methods may be used for allocating joint cost to the products.

- **Physical-Measures (or Simple-Average) Method**: This method uses a physical measure of the joint products, such as weight, count, volume, or package, that is common to all joint products at the split-off point.
- The major disadvantage of this method is that the physical measure may not have a relationship to the income-producing power of the products.
- **Weighted-Average Method**: This method is an improved version of the physical-measure method. This method assigns a weight factor based on the product's market value, labor time and skill needed, size, weight, etc.
- **Sales-Value at Split-off**: This method allocates joint costs to the separable products based on the sales value of each product at the split-off point. This method is simple and reasonable to apply when all joint products do not require further processing and are sold at the split-off point.

  Where joint products cannot be sold at the split-off point without incurring further processing costs, one of the following methods may be used for allocating joint costs to the products.
- **Relative-Market-Value (RMV)** or **Net-Realizable-Value (NRV)** Method: The RMV (or NRV) is the estimate of the sales value of each joint product, computed by deducting the separable cost from the potential final sales value for the product. Then, it uses the proportion of the approximated market value of each product relative to the total of the approximated market values of all products as the basis for joint cost allocation.
- **Constant-Gross-Margin (GM%) Method**: This method assumes that every joint product earns the same gross-margin ratio (GM%) as the basis for joint cost allocation. The same gross-margin ratio is computed in the following steps:

1) Compute the overall or total gross margin for all products combined, which is computed as follows:

$$TGM = TFSV - TJC - TSC$$

where
TGM = Total gross margin for all products combined;
TFSV = Total final sales value for all products combined;
TJC = Total joint cost;
TSC = Total separable cost.

2) Calculate the GMR (or GM%) for each product

$$GMR = \left(\frac{TGM}{TFSV}\right)$$

$$GM\% = 100 \times \left(\frac{TGM}{TFSV}\right)$$

3) Calculate the Joint Costs for each product as follows:

$$\text{Join cost allocated to a product} = GMR \times (\text{Final sales value of the product})$$

A **by-product** is a product having a relatively low sales value as compared to the main or joint products. By-products may be sold at split-off or processed further. However, if sales value of a by-product increases, it may be re-classified as a main product.

There are two commonly used methods of accounting to record the cost of the by-products. One method is the net realizable value of the by-product at the split-off point (final potential sales value minus separable cost). The second method is the zero-cost method by which no joint cost is allocated to the by-products; their only costs would be their required separable costs.

When the sales price varies with the output level for one or more joint products, the traditional method may not be capable of calculating proper joint cost allocation to the product. In such cases, a nonlinear programming can provide a viable solution. If there are constraints of resources (e.g., materials, capital, and production capacity), linear programming may be applied (which is not covered in this chapter).

A company should choose a method for joint cost allocation that is specifically appropriate to the values of its products and production requirements and consistently use the selected method.

REFERENCES

Kaplan, R. S., (1982). *Advanced Management Accounting*, (Englewood Cliffs, NJ: Prentice-Hall, Inc.)

Manes, R. P. and Smith, V. L., (1965). Economic Joint Cost Theory and Accounting Practice. *The Accounting Review*, XL(1): 31–35.

Myer, J. N., (1972). *Cost Accounting for Non-Accountants*, (New York, NY: Hawthorn Books, Inc.)

Needles, Jr., B. E.; Anderson, H. R. and Caldwell, J. C., (1993). *Principles of Accounting*, (5th edition, Boston, MA: Houghton Mifflin Company).

Tayyari, F. and Parsaei, H. R., (1992). Joint Costs Allocation to Multiple Products: Cost Accounting v. Engineering Techniques, Chapter 12 in: H. R. Parsaei and A. Mital (eds.): *Economic Aspects of Production and Manufacturing*, (London: Chapman & Hall).

Weil, Jr., R. L., (1968). Allocating Joint Costs, (Communications). *The American Economics Review*, LVIII(5, Part 1): 1342–1345.

Williams, D. J. and Kennedy, J. S., (1983). A Unique Procedure for Allocating Joint Costs from a Production Process? *Journal of Accounting Research*, XXI(2): 644–645.

REVIEW QUESTIONS

R5.1. What is a common cost?

R5.2. What is common cost allocation?

R5.3. Which joint cost allocation method is the most effective?

R5.4. What is the objective of joint cost allocation?

R5.5. Why would a number of accountants express a preference for the net-realizable-value method of joint cost allocation over the physical quantities method?

R5.6. When would a physical quantities method for allocation be preferred?

R5.7. What is the basic difference between the allocation of joint costs to (*a*) joint products and (*b*) by-products?

R5.8. What costs are irrelevant for the decision of whether to sell a joint product or process it further?

R5.9. On what basis should a company choose a joint cost allocation method?

R5.10. A mining company recovers gold, silver, copper, and lead by refining ore from the same mine site. What type of process would be this refining process?

R5.11. Which joint product allocation method produces the same gross profit percentage for all joint products?

R5.12. If a company produces different products using the production process or material, which of the method should it use to determine whether a product is the main product or a by-product?

PROBLEMS

P5.1. A small company manufactures ballpoint pens in four different colors: blue, black, green, and red. In a joint process, the company manufactures the outer housings, the cartridges (reservoirs), and tip points with balls of pens, which are the same in costs. However, the inks of different colors cost different and this cost is added to the individual pens separately. During a period, the company manufactures a total of 1,000,000 pens as follows:

Blue ink: 400,000 pens
Black ink: 300,000 pens
Green ink: 100,000 pens
Red ink: 200,000 pens

The total joint cost of the batch up to the split-off point is $150,000.

Required: Allocate the joint cost among all types of pens produced using the physical-measure (simple-average-unit-cost) method.

P5.2. Harold & Sons Company produces two joint products, A and B. The production process results in producing a by-product also. During the month of July of the year 20XX, 50,000 gallons of input materials were processed, which produced 30,000 gallons of product A, 15,500 gallons of product B, and 4,500 gallons of the by-product. The total joint cost of materials, labor, and overhead amounted $195,500. None of the products were salable at split-off point. Additional manufacturing (separable) processing costs were incurred as shown below:

Product	Separable Cost (per gallon)
A	$3.80
B	3.00
By-Product	1.00

Assume that during July all of the products were completed and sales were as follows:

Product	Sales Price	Sales Volume (gallons)
A	$13.00	25,000
B	11.00	12,500
By-Product	4.00	1,750

Assume, also, that the company uses the net realizable value of the by-product as a reduction in the production costs of the joint products.

Required

a) Use the Relative-market-value method to allocate the joint costs and compute the total manufacturing unit cost of each product.
b) Prepare a product-line income statement to analyze the profitability of the products.
c) Repeat part (a) using the physical-measure method to allocate the joint cost.

P5.3. A certain raw material costs a manufacturing company $30 per unit to process. Processing 1,000 units of this input material results in the production of 100 units of Product A, 250 units of Product B, and 500 units of a by-product. The following information is available:

	Product A ($)	Product B ($)	BP ($)
Separable Manufacturing Cost (per unit)	50	25	5
Sales Price	400	165	15

The company uses the "relative-market-value" method in its joint cost allocation and the "zero-cost" method to account for its by-products.

Required: Find the manufacturing unit cost for each product, including the by-product.

P5.4. Barnow Manufacturing Company produces two joint products, A and B. The production process results in producing a by-product also. During the month of July 20XX, 100,000 gallons of input materials were processed, which produced 45,000 gallons of Product A, 37,500 gallons of Product B, 17,000 gallons of a by-product, and 500 gallons of waste material. The total joint cost of materials, labor, and overhead amounted to $450,000. None of the products were salable at split-off point. Additional manufacturing (separable) processing costs were incurred as shown below:

Product	Separable Cost (per gallons)
A	$3.75
B	2.20
By-Product	1.00

Assume that during July, all of the products were completed and sales were as follows:

Product	Sales Price	Sales Volume (gallons)
A	$15.00	40,000
B	11.20	30,000
By-Product	2.00	15,000

Assume that the company applies no cost (zero cost) from the joint cost to the by-product.

Required

a) Use the relative-market-value method to allocate the joint costs and compute the total manufacturing unit cost of each product.

b) Prepare a product-line income statement to analyze the profitability of the products.

c) Repeat part (a) using the physical-measure method to allocate the joint cost.

P5.5. The Goodnext Company produces two products, A and B, from processing a common input raw material. One unit of the raw material yields 1 unit of Product A and 3 units of Product B. Assume that the supply of raw material is unlimited and it costs \$35 per unit. The variable cost of processing 1 unit of raw material is \$25 per unit. The Goodnext's economists have derived the following demand volume–price functions:

$$Q_A = 50 - P_A \text{ and } Q_B = 80 - 2P_B$$

where

Q_A = Number of units demanded of Product A;
Q_B = Number of units demanded of Product B;
P_A = Price of Product A;
P_B = Price of Product B.

Required: Find the optimal quantity of the raw material to be processed, the price of each product at the optimal sales volumes, and the amount of joint cost to be allocated to each product.

P5.6. The ABC Company produces 2 products, 1 and 2, from processing a common input raw material. One unit of the raw material yields 2 units of Product 1 and 3 units of Product 2. Assume that the supply of raw material is unlimited and it costs \$25 per unit. The variable cost of processing 1 unit of the raw material is \$20 per unit. The ABC's economists have derived the following demand volume–price functions:

$$Q_1 = 47.5 - P_1 \text{ and } Q_2 = 80 - 2P_2$$

where Q_1 = Sales volume of Product 1;
Q_2 = Sales volume of Product 2;
P_1 = Sales price of Product 1;
P_2 = Sales price of Product 2.

Required: Find the optimal quantity of the raw material to be processed, the price of each product (i.e., P_1 and P_2) at the optimal sales volumes (i.e., Q_1 and Q_2), and the amount of joint cost to be allocated to each product.

P5.7. By processing a certain input material, a company produces three products (e.g., Product 1, Product 2, and Product 3) in fixed proportions. Processing each unit of the input material results in producing 0.2 unit of

Product 1, 0.1 unit of Product 2, and 0.4 unit of Product 3. Assume that the supply of the material is not limited and costs $50 per unit. A market study has indicated the following weekly demand–price functions:

$$Q_1 = 400 - 4P_1, Q_2 = 60 - 0.2P_2 \text{ and } Q_3 200 - 4P_3$$

or inversely

$$P_1 = 100 - 0.25Q_1, P_2 = 300 - 5Q_2 \text{ and } P_3 = 50 - 0.25Q_3$$

where Q_1 = Demand quantity for Product 1;
Q_2 = Demand quantity for Product 2;
Q_3 = Demand quantity for Product 3;
P_1 = Sales price for Product 1;
P_2 = Sales price for Product 2;
P_3 = Sales price for Product 3.

Required: What quantity of the input material should be processed to maximize the company's profit? Find the amount of joint cost to be allocated to each unit of the products, the optimal sales quantity, and sales price for each product.

P5.8. Three different products (namely, Product 1, Product 2, and Product 3) are, simultaneously, produced through a common process in fixed proportions. Processing each unit of the input material results in the production of:

1.0 unit of Product 1;
0.5 units of Product 2;
1.5 units of Product 3.

Assume that the supply of the material is not limited and costs $50 per unit and that it costs an additional $60 per unit to process the input material. A market study has indicated the following weekly price–demand relationships:

$$Q_1 = 200 - 0.8P_1,$$

$$Q_2 = 40 - 0.4P_2, \text{ and}$$

$$Q_3 = \frac{800}{3} - \frac{2}{3}P_3$$

where Q_1 = Demand quantity for Product 1, when its sales price is P_1;
Q_2 = Demand quantity for Product 2, when its sales price is P_2;
Q_3 = Demand quantity for Product 3, when its sales price is P_3.

Required: What quantity of the input material should be processed to maximize the company's profit? Find the amount of joint cost that should be allocated to each unit of each product, also the optimal sales quantity and sales price for each product.

P5.9. A certain raw material costs a manufacturing company $34.50 per unit to process. Processing 20,000 units of this input material results in the production of 5,000 units of Product A, 11,500 units of Product B, and 10,000 units of a by-product. The following information is available:

	Product A ($)	Product B ($)	By-Product ($)
Separable Manufacturing Cost (per unit)	30	10	2.50
Sales Price	200	110	4.50

The company uses the "relative-market-value" method in its joint cost allocation and the "zero-cost" method to account for its by-products.

Required: Find the manufacturing unit cost for each product, including the by-product.

6 Manufacturing Cost Estimation

LEARNING OBJECTIVES

- Understand the concept of and needs for cost estimation.
- Understand reasons for estimating fixed and variable costs.
- Estimate costs using the account analysis method.
- Estimate costs using scatter-graph method.
- Estimate costs using the high-low method.
- Estimate costs using the linear regression method.
- Perform regression analysis using Excel® Solver software to fit regression line into cost data.
- Interpret the results of linear regression analysis.

6.1. INTRODUCTION

Manufacturing cost estimation is a cost analysis process for predicting the cost of manufacturing a product. Whereas, product costing, as presented in the previous chapter, is a systematic procedure for accurately recording all allocated costs to determine the cost of a manufactured product.

The cost estimation is imperative in managerial decision-making (such as which product lines will be profitable and which product lines need to be reengineered to stay cost-competitive). Cost estimation helps managers in making more successful decisions in making their manufacturing systems cost-effective.

The depth of this analysis ranges from some simple calculations to a complex mathematical and statistical methodology.

In a general sense, estimation is the prediction or forecast of resources (time, cost, and materials) required to achieve or obtain an agreed-upon scope (of a project or a volume of a product). Cost estimation is a process that helps to identify resource requirements and capabilities, setting priorities, and allocating resources for project planning and controlling. Costing and valuation are the byproducts of cost estimation and help decision-makers in assessing their decision alternatives (Yeung, 2018).

The usefulness of estimation depends on its accuracy, which is judged by the degree to which the final cost outcome for a given project will vary from the original estimated cost. Nevertheless, there should be some allowance for contingencies, which is an amount added to an estimate to allow for items, conditions, or events, such as inflation and uncertainty, that will likely result in additional costs.

A detailed cost estimating process determines the overall cost estimate by summing detailed estimates done at lower levels of the ***Work Breakdown Structure***

DOI: 10.1201/9780429432163-6

(WBS). The WBS is a technique that breaks down the system being costed into lower-level components (such as parts or assemblies), each of which is costed separately for direct material, direct labor, and overhead costs. Estimates of direct materials are often based on the bills of materials, which list the number of units and the unit costs of all components that go into each product. Estimates for direct labor hours are usually based on analyses of engineering drawings and some contractor or industry-wide standards. Finally, the estimates for the overhead costs based on predetermined rates are described in Chapter 3.

6.2. GENERAL TYPES OF COST ESTIMATES

For planning and determining resource requirements (budgeting) for any projects, decision-makers need to have cost estimation. As the word implies, estimates are not definite. The accuracy of an estimate depends on the project stage and the reliability of data sources. The following are the three types of cost estimation, presented in the order of their increasing level of accuracy (Hassib, 2012; Yeung, 2018):

1) Order of magnitude (ROM) estimate.
2) Budget estimate.
3) Definitive estimate.

- **Order of magnitude (ROM) estimate:** A ROM estimate is a rough estimate of costs used at the very early stage of a project, particularly during the planning stages. This type of estimation provides a "ball-park" estimate based on the information available at the time for the purpose of having an idea about general and total expenditures instead of itemizing costs based on the project activities.

  Planners usually use historical data of similar projects or an aggregate of data from different projects to perform an order of magnitude estimation. This type of estimate has a range of variance from –25% (underestimate) all the way to +75% (overestimate) of the actual costs; this range can become as wide as 50% underestimation to +100% overestimation (Hassib, 2012; Yeung, 2018; Phillips, 2020).
- **Budget estimate**: A budget estimate (or **top-down** estimate) is a preliminary cost estimate, which is more accurate than an order of magnitude estimate. The purpose of this type of estimation is to provide a preliminary itemized list of expenses based on the main components of the project. With the budget estimate, the estimator starts at the top of the project (major components) and works through to the bottom of it (details). The accuracy of this type of estimate is from 10% underestimate to 25% overestimate (Hassib, 2012; Yeung, 2018; Phillips, 2020). Therefore, this is done at the planning stage of the project. Estimators, usually, make this type of estimation using historical data of a related project or obtainable data from the project plan, the expected labor cost, and costs of required equipment and materials.

- **Definitive estimate:** A definitive estimate (also known as detailed cost estimate or bottom-up estimate) is the most accurate of the three types of cost estimation. This type of estimation breaks down the whole project into its detailed compositions to prepare a list of all requirements, to each of which an estimated cost is assigned. Definitive estimates have an accuracy of about 5% underestimate to 15% overestimate (Hassib, 2012; Yeung, 2018).

  With any type of estimate, the estimator should provide the range of variance in the costs and an explanation of how he/she made the estimate. Otherwise, the customer may take the quoted cost as a promised cost (Phillips, 2020).

6.3. ASSUMPTIONS FOR THE USE OF COST ESTIMATING MODELS

Cost estimates are usually based on limited information and under certain constraints. These constraints usually take the form of assumptions that bind the estimate's scope, establishing baseline conditions under which the estimate will be made. The following are three important assumptions that must be considered when estimating costs using the methods presented in this chapter.

1) The variable costs per unit and total fixed costs assumed will remain constant at all the activity levels.
2) The cost to estimate is assumed to have a linear relationship with the activity level (production quantity).
3) When costs are estimated for a planned activity level, it is assumed that the activity level is within the relevant range.

These assumptions are realistic as long as the relevant range is clearly identified and the linearity assumption does not significantly distort the resulting cost estimate.

In cost analysis, the term ***relevant range*** refers to a normal range of activity levels in which the total fixed costs will not change as the volume of activity changes. However, if a firm's activity level drops to an extremely low level, the management may take action to decrease the fixed costs. Similarly, if the firm's activity level increases beyond the relevant range, the firm would likely incur more fixed costs. Then, the linearity assumption for the cost behavior as a straight line (linear) may become reasonable.

Cost analysts should analyze costs within a relevant range to arrive at a valid cost estimate. The range for a projection is usually between the upper and lower limits (bounds) of past activity levels for which data is available.

6.4. METHODS FOR DEVELOPING A COST ESTIMATING MODEL

Cost estimation is the process of developing an approximation of the total cost of a product, program, project, or operation. Cost estimation is crucial in managerial decision-making and budgeting.

For manufacturing product costing and profitability analysis, cost estimators establish ***cost estimate relationship*** (CER), which may be an algorithm or formula that is used to perform the costing. A cost estimate function relates cost with quantities such as time and quantity or some other relevant elements.

There are various techniques for estimating the future costs of a production system. The use of a specific approach will depend on the purpose of estimation and the amount and type of available data.

In many production decision-making cases, the cost analysis involves the estimation of two types of manufacturing costs: fixed costs and variable costs. For producing a product, labor and material costs are usually variable costs that vary as the production output changes. A manufacturer consumes more labor and materials to produce more units of its product. Accordingly, the cost of labor and material varies in a direct proportion to the level of output.

To estimate the total cost associated with planned production in a future period, analysts develop a mathematical cost equation. To do this, they must first determine the amounts of variable and fixed costs associated with producing their products. The cost equation for a given product is typically the direct (variable) cost of manufacturing per unit of the product multiplied by the number of units produced and sold plus the fixed (overhead) costs. For example, a company with \$100,000 in overhead costs that makes products that cost the company \$6 each to manufacture and sell has a cost equation of \$100,000 + (\$6 × number of units sold). We can generalize this cost equation in the following form:

$$Y = a + bX \tag{6.1}$$

or

$$C = F + VQ \tag{6.2}$$

where

$Y = C$ = Total (mixed) cost;
$a = F$ = Total fixed cost;
$b = V$ = Variable cost per unit;
$X = Q$ = Measure of activity or number of units produced;
$bX = VQ$ = Total variable cost.

A company can use its cost estimating equations to determine how much would it cost to make a certain level of sales. The equation would also allow the company to determine the price it must charge for a certain number of units sold in order to break even. Once the company has developed its cost equation and determined its product's selling price, the company can construct a profit equation, called the Cost-Volume-Profit (CVP) model. With the CVP model, the company can perform a variety of critical analyses, which will be presented in Chapter 7.

There are four common approaches to identify the production cost behavior, as fixed and variable costs, and construct the cost estimation equation:

- Account analysis.
- High-low method.
- Scatter-graph method.
- Regression analysis.

6.4.1. Account Analysis Method

The account analysis approach may be the most common starting point for estimating fixed and variable costs. This approach requires basic knowledge of accounting and experience in cost analysis to review the relevant accounts to obtain cost data and determine whether each cost item is fixed or variable. The total of all costs identified as fixed gives the estimate of total fixed costs. Similarly, the total of all costs identified as variable provides an estimate of total variable costs, when divided by the measure of activity (i.e., number of *units produced*), the result would be variable cost per unit.

Example 6.1

The cost analyst at the Sambo Scooter Manufacturer asked the accounting department to provide costs data regarding the scooters' production department for the month of August. The accountants provided the following financial information about the scooters production department in August:

Number of scooters produced and sold	10,000
Materials used	$80,000
The labor used in scooter department:	
Assembly labor cost	14,500
Supervisory cost	5,000
Facility costs:	
Rent	3,250
Utilities	750
Insurance	500
Total Production Cost	**$60,000**

The analyst carefully reviewed the costs data in consultation with the production manager. They sorted out the variable and fixed costs for the month of August as follows:

	Fixed	Variable	Total
Materials used	$0	$80,000	$80,000
Labor used in scooter department			
Assembly labor cost	0	14,500	14,500
Supervisory cost	5,000	0	5,000
Facility costs			
Rent	3,250	0	3,250
Utilities	250	500	750
Insurance	500		500
Total Production Cost	**$9,000**	**$95,000**	**$104,000**

The estimated total fixed cost (F) is $9,000, and the variable cost per unit (rate) equals the total variable cost of $95,000 divided by 10,000 units produced in August, as computed below:

$$V = \$95{,}000/10{,}000 \text{ units} = \$9.50 \text{ per unit}$$

The analyst would present the total cost (C) equation for estimating the future monthly total cost, given the planned monthly production level (Q), as follows:

$$C = \$9{,}000 + \$9.50Q$$

For example, if the manufacturer plans to produce 12,000 scooters in the following month, its estimated total production cost would be:

$$C = \$9{,}000 + \$9.50(12{,}000) = \$123{,}000$$

However, this estimate is reasonable if the level of production (12,000 units) is within the relevant range, which does not require any additional resources.

In addition, one must be mindful when using the historical data for only one period (month) to estimate future costs, as that specific period may not be a typical period for the demand for the product and production costs. For example, when demand is higher in the months of May through August, the company may get quantity discount on materials, the utility costs may be low relative to those in the winter months, and other production costs could significantly change as the company may make different arrangements for handling increased or decreased demand during different months. To average out these fluctuations, one should use the historical data over a year or more to develop a cost estimating model.

The major advantage of the account analysis method is that managers and accountants are familiar with the company operations and how costs incur as the activity level changes. However, this method has some disadvantages. Managers and accountants may apply biased judgments; also, imperfect decisions often have major economic consequences for the company.

6.4.2. Scatter-Graph Method

The scatter-graph method is a visual technique used for estimating the fixed and variable elements of a semi-variable cost (i.e., a mixed cost) to budget for future costs. Using the scatter-graph method, cost analysts plot the cost against the level of activity on a graph and draw a line that they think is the best-fit line through the points.

A scatter-graph uses a horizontal axis that represents an activity level (e.g., production quantity, labor hours, machine hours) and a vertical axis that represents its cost. Pairs of data points (i.e., activity levels and their corresponding costs) are plotted on the graph and a regression line that runs through the dots represents the best fit of the relationship between the variables. After determining the best-fit line through the data points, the fixed cost is found at the intercept, the point where the regression line crosses the vertical axis (y-axis), and the slope of the line is the variable cost per unit. The procedure requires the following four steps:

Step 1: Draw a Scatter-Graph: Plot the coordinates of data points (the activity levels and their corresponding total costs) on graph paper, the activity levels (i.e., production units, labor hours, machine hours, etc.) along the horizontal axis (x-axis), and total mixed cost along the vertical axis (y-axis).

Step 2: Draw a Regression Line: Draw a regression line through the scattered data points by visual inspection and try to minimize the total vertical distances between the line and all the data points.

Step 3: Find Total Fixed Cost (*F*): Extend the regression line toward the y-axis to locate its intercept; that is, the point at which the regression line crosses the y-axis. Total fixed equals the y-intercept of the regression line.

Step 4: Compute Variable Cost Per Unit (*V*): Variable cost per unit is equal to the slope of the line. Take the coordinates of two points (X_1, Y_1) and (X_2, Y_2) on the regression line and compute the variable cost per unit using the following formula:

$$\text{Variable Cost per Unit } (V) = \text{Slope of Regression Line}$$

$$= \frac{\text{Change in Total Cost}}{\text{Change in Activity Level}}$$

The symbolic form of the formula is $V = \dfrac{Y_2 - Y_1}{X_2 - X_1}$

Since the best-fit line is drawn subjectively, different individuals have different ideas about drawing the line. That is, each person would have a somewhat different estimate of the cost.

Example 6.2

Table 6.1 represents the monthly production cost information for Meemee Co. during the previous year. Using the scatter-graph method, derive a cost estimating equation for the company.

Solution

Step 1. Plot the data points for each period (month) on a graph (see Figure 6.1).

Step 2. Visually fit a line through the data points such that to minimize the distance from the data points to the line (i.e., to make the line as close to the data points as possible). Make sure the line touches at least one data point. Extend the line to the vertical axis. This is the total production cost curve. Figure 6.1 shows the line through the data points. Notice that the line hits the data point for 2,800 units produced and $220,000 total cost (February production and cost data).

TABLE 6.1
Monthly Production Costs for Meemee Co.

Month	Units	Costs	Month	Units	Costs
January	**2,200**	**$185,000**	July	4,200	$285,000
February	2,800	220,000	August	5,300	315,000
March	3,500	250,000	September	4,500	295,000
April	3,800	235,000	October	3,500	210,000
May	**6,100**	**365,000**	November	3,000	210,000
June	5,200	280,000	December	2,600	170,000

*Bold highlights the data points to be used.
Bold highlighted data points are the high (6,100) and low (2,200) volume points.

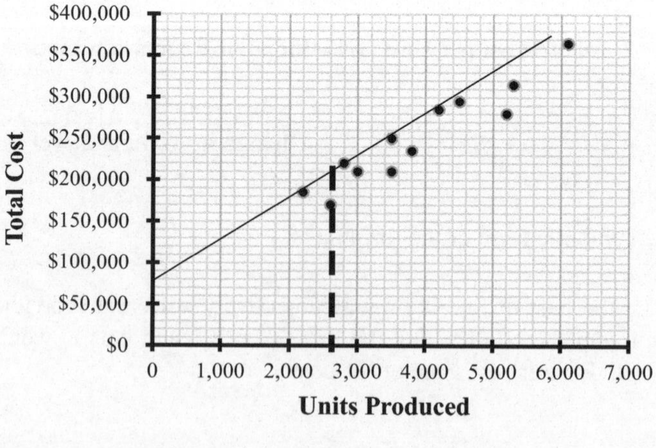

FIGURE 6.1 Scatter-graph of total production costs for Meemee Company (Example 6.2).

Step 3. Estimate the total fixed cost (*F*). The total fixed cost is simply the point at which the total cost line drawn in Step 2 crosses the vertical axis. This is known as the *y*-intercept. Recall that the fixed costs remain the same in total regardless of the level of production and variable costs change in total with changes in levels of activity. Where the total cost line intercepts the *y*-axis (total cost axis), the total cost equals the fixed cost (plus zero variable cost), which corresponds to 0-activity (or production) level. Therefore, the fixed cost for Meemee Co. is approximately $55,000. The equation for the total cost line drawn in Step 1 is partially complete and stated as:

$$C = \$55,000 + VQ.$$

Step 4. Calculated the variable cost per unit (*V*). Fill the data point through which the line passes, that is, $Q = 2,800$ units and $C = \$220,000$, in the partially completed equation in Step 3 to find the unknown variable cost per unit, as follows:

$$C = F + VQ$$

$$\$220,000 = \$55,000 + V(2,800)$$

$$V = \$58.93 \text{ (rounded)}$$

The variable cost is approximately $58.93 per unit.

Step 5. State the results in equation form $C = F + VQ$. We found in Step 3 that the total fixed cost is $55,000, and in Step 4 that the variable cost per unit is $58.93. Now, we can state the total cost estimating equation as:

$$C = \$55,000 + \$58.93Q.$$

Now, the company can estimate total production costs for a given level of production (*Q*). For example, if the company plans to produce 5,500 units during an upcoming month, the estimated total production cost is found as follows:

$$C = \$55,000 + \$58.93(5,500) = \$379,115.$$

6.4.3. High-Low Method

The high-low method provides is an easy and quick approach to estimate fixed and variable costs, which may not be as accurate as other methods. It uses the historical data during several reporting periods; but, it uses only the data points corresponding to the **highest** and **lowest** levels of activity (output) to derive the algebraic equation for the total cost. Hence, it ignores all the data other than the two extreme data. The slope of the resulting equation is the variable cost and its *y*-intercept is the fixed cost.

However, if the highest and lowest levels of costs are not representative of the costs to estimate, then this method would either overestimate or underestimate the costs.

Given a historical production and total cost data set from several reporting periods, the high-low method uses the following steps to develop the total production costs equation of the form $C = F + VQ$:

Step 1. Identify the highest and lowest levels of activity (units produced) from the given data set.

Step 2. Calculate the variable cost per unit (V). Use the following formula to find V:

$$V = \frac{C_H - C_L}{Q_H - Q_L} = \frac{\text{Change in Total Cost}}{\text{Change in Activity Level}} \tag{6.3}$$

where

C_H = Total cost at the highest level of activity (production quantity);
C_L = Total cost at the lowest level of activity;
Q_H = The highest level of activity (production quantity);
Q_L = The lowest level of activity;
V = Variable cost per unit.

Step 3. Calculate the total fixed cost (F). Fill the units cost found in Step 1 (as the slope) and the data point for either the lowest or highest activity levels (as the coordinates of a point on the total cost curve) in the total cost equation of form $C = F + VQ$, as follows:

$$C_L = F + VQ_L$$

or

$$C_H = F + VQ_H$$

Then, solve the equation for F to find the estimated value for the total fixed cost:

$$F = C_L - VQ_L$$

or

$$F = C_H - VQ_H$$

Step 4. State the results in equation form $C = F + VQ$. Replace F and V in the total cost equation with the computed variable cost per unit (V) found in Step 2 and the fixed cost (F) found in Step 3 to find the total cost estimating equation for the problem at hand.

The following example clarifies the steps taken in the high-low method of cost estimation.

Example 6.3

Use the monthly production costs data for Meemee Co., as given in Example 6.2, to determine the total cost estimating equation using the "high-low" method.

Solution

Step 1. Although a graph isnot required in the high-low method, it is a helpful visual toolto identify data points, one with the **highest** level of activity and the other with the**lowest** level of activity.Alternatively, we can identify the two extreme data points byarranging the data points in an ascending or descending order ofthe activity levels (units produced). The circled data pointsshown in Figure 6.2 are the required ones.

The lowest level of production was 2,200units (in January) and the highest level of production was 6,100 units(in May). The difference between the number of units produced and thedifference between the total cost at the highest and lowest levels ofproduction are shown below:

	Production (units)	Total Cost ($)
Highest level	6,100	365,000
Lowest level	2,200	185,000
Difference	**3,900**	**180,000**

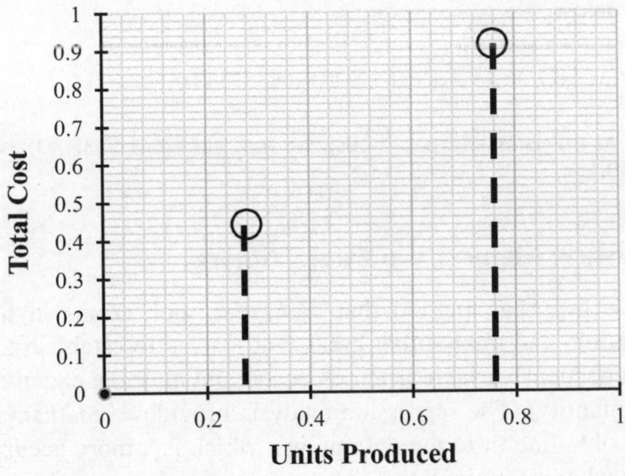

FIGURE 6.2 The circled data points are the high- and low-activity levels in the production and cost datafor Meemee Co. (Example 6.3).

Step 2. Calculating thevariable cost per unit (*V*):

$$V = \frac{C_H - C_L}{Q_H - Q_L} = \frac{\$365{,}000 - \$170{,}000}{6{,}100 - 2{,}200} = \frac{\$180{,}000}{3{,}900} = \$46.15 \text{ (rounded)}$$

Step 3. Calculating the totalfixed cost (*F*). Simply useeither the data for the high- or low-activity level and fill inthe data in the equation and solve for *F* (total fixed costs). For example, let's plug the data for the high activity level (6,100 units,$365,000) into the equation, as follows:

$$\$365{,}000 = F + 46.15(6{,}100)$$

Then, solving for *F* gives the estimated total fixed cost, as $F = \$83{,}485$.

We will get the same result using the data for thelow activity level (2,200 units, $185,000):

$$\$185{,}000 = F + 46.15(2{,}200)$$

Then, solving for *F* gives the estimated total fixed cost as $F = \$83{,}470$. The differencebetween the two values is due to the rounding effect used for thevariable cost per unit. However, as this is an estimated cost, we mayeven state it as $83,500.

Step 4. Stating the resultsin a specific equation for the problem as:

$$C = \$83{,}485 + 46.15Q.$$

Now, if the company plans to produce 5,500units during the upcoming month, its estimated total production costswill be $386,000, as computed below:

$$C = \$83{,}485 + 46.15(5{,}500) = \$337{,}310.$$

Again, as this is an estimated cost, we may stateit as $337,300 or even as $337,000.

6.4.4. Regression Method and Excel® Solver

Regression is a statistical method that has great applications in finance, investing, cost estimation, and many other areas that is used to establish the relationship between one dependent variable (such as costs) and an independent variable (such as production quantity). The regression analysis provides a mathematical equation for the best possible line fit to the data points, which is a more accurate future cost estimation than what is obtained through a scatter-graph.

Regression analysis is a statistical method for estimating the relationships between one or more independent variables and dependent variables. The equation for a line is $Y=a+bX+\varepsilon$. In this equation, Y is the dependent variable whose future value we want to predict when the independent variable X will be of a certain value, a is the y-intercept, b is the slope or rate of change in Y per unit change in X, and ε (epsilon) is the residual (error) of the estimate. The formulas to calculate the intercept "a" and the slope "b" are as follows:

$$b=\frac{n(\Sigma XY)-(\Sigma X)(\Sigma Y)}{n(\Sigma X^2)-(\Sigma X)^2} \tag{6.4}$$

$$a=\bar{Y}-b\bar{X} \tag{6.5}$$

Based on the symbols used in this chapter for fixed, variable, and total costs and the activity (production) level, we set the following correspondence between the two sets of symbols:

$a=F=$Fixed cost (intercept);
$b=V=$variable cost per unit (slope);
$n=$Number of observations (number of reporting periods);
$X=Q=$Activity level (independent variable);
$Y=C=$Total cost (dependent variable);
$\bar{X}=\bar{Q}=\sum X/n=$Average of activity levels (average of production quantity);
$\bar{Y}=\bar{C}=\sum C/n=$Average total cost.

One of the greatest benefits of the regression method is its level of accuracy as compared to the scatter-graph and high-low methods. The scatter-graph method of cost estimation is somewhat subjective as the analyst draws his or her best visual fit line through the cost data points. The high-low method only uses the highest and lowest activity levels for cost estimation. Therefore, if these two data points are not representative of the true cost behavior, the estimation would be biased.

The major disadvantage of the regression method is the difficulty in manual calculation of the equation's slope and intercept estimates. The use of computer software, such as spreadsheet software, helps overcome the level of difficulty in calculations. If the software is not available, an understanding of algebra and some knowledge of statistics are required to perform computations. However, since the process is computationally intensive, there are possibilities for computational errors that would affect the accuracy of the estimates. As such, in the case of unavailability of computer software, the estimators may use another method.

Example 6.4

Using the regression method, develop the equation for estimating the total monthly production cost for Meemee Co. based on its historical production costs given in Table 6.1 of Example 6.2.

Solution

We add two more columns to the table of monthly production costs for Meemee Co. to perform the preliminary computations for components of the regression equations (6.4 and 6.5), as shown below:

Month	Units (X)	Costs (Y)	X^2	X.Y
January	2,200	$185,000	4,840,000	407,000,000
February	2,800	220,000	7,840,000	616,000,000
March	3,500	250,000	12,250,000	875,000,000
April	3,800	235,000	14,440,000	893,000,000
May	6,100	365,000	37,210,000	2,226,500,000
June	5,200	280,000	27,040,000	1,456,000,000
July	4,200	285,000	17,640,000	1,197,000,000
August	5,300	315,000	28,090,000	1,669,500,000
September	4,500	295,000	20,250,000	1,327,500,000
October	3,500	210,000	12,250,000	735,000,000
November	3,000	210,000	9,000,000	630,000,000
December	2,600	170,000	6,760,000	442,000,000
Total	46,700	3,020,000	197,610,000	12,474,500,000
Symbol	$\sum X$	$\sum Y$	$\sum X^2$	$\sum XY$

Now, we can use the summation terms in the regression equations, as follows:

$$b = \frac{n(\sum XY) - (\sum X)(\sum Y)}{n(\sum X^2) - (\sum X)^2}$$

$$= \frac{12(12{,}474{,}500{,}000) - (46{,}700)(3{,}020{,}000)}{12(197{,}610{,}000) - (46{,}700)^2} = 45.476$$

$$a = \bar{Y} - b\bar{X} = (3{,}020{,}000/12) - 45.476(46{,}700/12) = 74{,}689$$

The total cost estimating equation will then be:

$$C = \$74{,}689 + \$45.476Q$$

The cost estimators can use computer software, such as Microsoft Excel® Solver, Minitab®, and SAS®, to perform a regression analysis. The following is the instruction for using the Microsoft Excel® Solver add-in.

Using Excel® Solver [You need to activate the Solver in Excel®]

1) Enter the data set in the Excel spreadsheet, output level and total cost in two columns, side-by-side.

2) Click on the "Data" button at the center-top of the window.
3) Click "Data Analysis" (of Solver)
4) Choose the "Regression" option, and click OK.
5) In the new window:
    - Choosing the "Input Y Range" $C$2:$C$14, for example, chooses data in cells C2 through C14.
    - Choosing the "Input X Range" $B$2:$B$14, for example, chooses data in cells B2 through B14.
    - Output Range: Click on a cell below your data set, for example, $A$17 (for choosing cell A17)
6) Click on OK to see the results of the regression analysis.

For the previous example, using the regression analysis of the Excel Solver® program, we get $F = \$74,689$ and $V = \$45.476$ per unit. The program generated the output shown in Figure 6.3.

We also get $R^2 = 0.89873$, which is a measure of goodness of fit for linear regression models. This statistic indicates about 90% of the variability in the dependent variable (total cost in this example) is explained by the independent variables (output quantity in this example). The cost estimation model, as found manually, is:

$$C = \$74,689 + \$45.476Q$$

where the fixed cost is $74,689, which may be set as $74,700 or even $75,000, and the variable cost is $45.476, which may be set as $45.48, per unit. Figure 6.4 shows the regression line fit into the given data in our example.

SUMMARY OUTPUT of Excel Solver

Regression Statistics	
Multiple R	0.94801
R Square	**0.89873**
Adjusted R Square	0.88860
Standard Error	19230.5
Observations	12

ANOVA

	df	*SS*	*MS*	*F*	*P-value*
Regression	1	32,818,533,494	32,818,533,494	88.74351	0.00000274
Residual	10	3,698,133,172	369,813,317		
Total	11	36,516,666,667			

	Coefficients	*Standard Error*	*t Stat*	*P-value*	*Lower 95%*	*Upper 95%*
Intercept	**74,689**	19,589.71	3.812671	0.00341425	31,040.53	118,337.72
X (or Q)	**45.4760**	4.82741	9.42038	0.00000274	34.720	56.232

FIGURE 6.3 Excel® Solver output regarding regression line for data in Example 6.4.

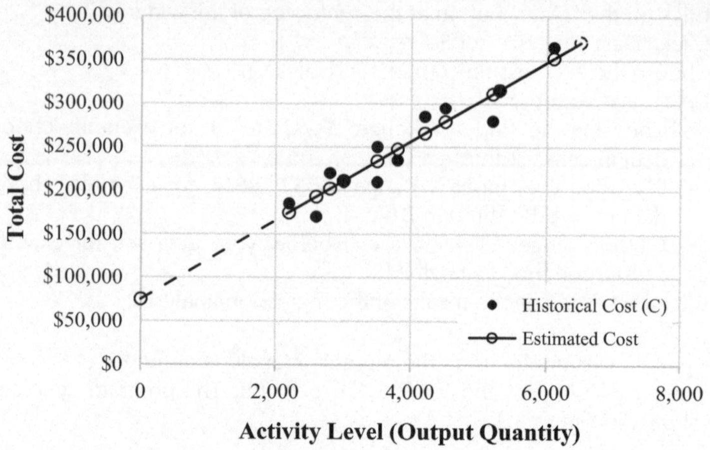

FIGURE 6.4 The regression line that fits into the cost-output data in Example 6.4.

SUMMARY

A cost-volume-profit (breakeven) analysis (Chapter 7) for a product(s) is only possible when the product's sales price and variable and fixed costs are known. This chapter presented four commonly practiced methods for establishing the relationship between production output and total cost to predict future (e.g., upcoming) period fixed cost and the variable cost per unit of the planned output level.

The ***account analysis method*** requires that a knowledgeable cost analyst, perhaps in consultation with other individuals, determine and sort out the costs in terms of fixed, variable, or mixed. If such a person is not available, another method should be used.

The ***high-low method*** uses only the highest and lowest activity levels and performs four steps to estimate fixed and variable costs: (1) identifying the high- and low-activity data points, (2) computing variable cost per unit, (3) computing the total fixed cost, and (4) stating the cost estimation equation.

The ***scatter-graph method*** uses five steps: (1) plotting all points on a graph, (2) drawing a line that best fits through the points with the line passing through at least one data point and extended to intercept the *y*-axis (total cost axis), (3) choosing the *y*-intercept as the total fixed cost (F), (4) plugging this fixed cost value and the coordinates of the point through which the line passes in the equation $C = F + VQ$ to find the variable cost per unit (V), and (5) stating the specific cost estimating equation using the computed F and V. The scatter-graph is also used to identify any outliers or data points that are unusually apart from the others, which may represent some unusual circumstances in the historical data set.

The ***regression method*** mathematically determines the equation for the total cost curve (line) that best fits the data points. Software packages such as Excel Solver® are available to perform regression analysis.

The objective of all four methods is to develop the cost estimating equation of the form $C = F + VQ$. The regression analysis provides the most accurate method as it provides a cost equation that best fits a line through all the data points. However, the goal of most analyses is to get a close predictive model, not necessarily a perfect one.

By now, you have learned that each method has its own advantages and disadvantages and the choice of a method will depend on the situation at hand. Therefore, the following should be considered in choosing a method for developing a model for predicting the total cost for a planned production output:

- Experienced employees may be able to effectively estimate fixed and variable costs by using the account analysis approach.
- If a quick estimate is needed, the high-low method may be the choice.
- The scatter-graph method is used to identify unusual data points (outliers), which may be excluded in developing a predictive cost model.
- Regression analysis is the most accurate method for developing a predictive model for manufacturing cost estimation; the use of computer software such as Excel® Solver makes it, probably, the most efficient method and the easiest to use.

REFERENCES

Hassib, A., (2012). How to Develop a Rough Order of Magnitude Estimate (ROM Estimate). *PM Documents, Project Management Made Simple*. Retrieved April 10, 2021 from: https://pmdocuments.com/how-to-develop-a-rough-order-of-magnitude-estimate-rom-estimate/ (posted: August 20, 2012).

Phillips, J., (2020). Project Cost Management. *Project Smart*. Retrieved April 10, 2021 from: https://www.projectsmart.co.uk/project-cost-management.php.

Yeung, N., (2018). Three Types of Cost Estimation in Projects. Retrieved April 10, 2021 from: https://www.profolus.com/topics/three-types-of-cost-estimation/ (Posted: March 15, 2018).

REVIEW QUESTIONS

R6.1. What is estimation?

R6.2. What is accuracy?

R6.3. What is contingency in cost estimation?

R6.4. What is Cost Estimate Relationship (CER)?

R6.5. On a scatter-graph for cost and output level, at what point the regression line intercepts the *y*-axis?

R6.6. How is the account analysis approach used to estimate fixed and variable costs?

R6.7. How can one use the high-low method to estimate fixed and variable costs?

R6.8. Which cost estimation is used to classify cost accounts as fixed or variable with respect to specific output levels?

R6.9. In what situation is the slope coefficient of a cost function equal to zero?

PROBLEMS

P6.1. The Albina Products Company's production executive officer desires to learn the behavior of production costs for a previous month, in which the company produced 7,500 units of a given product. She asked the production manager to analyze the production cost accounts for that month and classify the costs as fixed and variable. The production manager provided the following information:

	Total	Variable	Fixed
Materials used	$480,000	$480,000	$ 0
Labor used in	370,000	270,000	100,000
Factory facility costs	230,000	60,000	170,000
Total Production Cost	**$1,080,000**	**$810,000**	**$270,000**

Required

a) Express the production costs by a mathematical equation of the form $C = F + VQ$.

b) Assume that the company has planned to produce 8,000 units next month. Estimate the total production costs for the month.

P6.2. Zaloo Company is using the account analysis method to identify the behavior of production costs during the month of June, in which it produced 7,000 units. The production manager analyzed the production costs in June and reported the following information to the company's management:

	Variable	Fixed	Total
Materials used in production	$244,000	$ 0	$244,000
Labor used in production	96,000	12,000	108,000
Factory overhead costs	68,000	150,000	218,000
Total product cost	**$408,000**	**$162,000**	**$570,000**

Required

a) Express the production costs in an equation of $C = F + VQ$ form.

b) Assume Zaloo Company has planned to produce 8,000 units in the month of July. Estimate the total production costs for that month.

P6.3. For each of the following three per-unit costs, determine whether it is a variable, fixed, or mixed cost:

Production (units)	Cost A ($)	Cost B ($)	Cost C ($)
10,000	10.00	12.00	15.00
20,000	10.00	8.50	7.50

Required: What type of cost is each of the three costs?

P6.4. A manufacturing firm uses the high-low method of estimating costs. The firm had total costs of $45,000 at its lowest level of activity, when 5,000 units were sold. When, at its highest level of activity, sales equaled 10,000 units, and total costs were $70,000.

Required: Develop a cost estimation model for the company using the high-low method.

P6.5. ZTF Corporation accumulated following data for one of its production activities:

Output Level (units)	Total Costs ($)
13,800	128,000
14,700	131,000
15,500	136,700
12,600	120,900
10,500	109,700

Required

a) Derive a total cost estimating equation using the high-low method based on given data.

b) Compute the predicted total cost if the output level were 15,000 units.

P6.6. KTG Company accumulated the following data for a delivery truck during the first four months of the year.

	Distance Driven (miles)	Total Cost ($)
January	8,000	11,100
February	9,000	11,600
March	10,500	12,500
April	11,500	13,550

Required

a) Develop the equation to predict total costs for the delivery truck using the high-low method.

b) Estimate the total costs of driving the truck for 12,000 miles.

P6.7. Data concerning an operation activity for the first six months of the year are as follows:

	Machine Hours	Electrical Cost ($)
January	5,000	5,250
February	5,200	5,360
March	4,900	5,195
April	5,600	5,595
May	5,200	5,400
June	6,500	6,075

Required: Develop a total cost estimating based on the given data using:

a) The **high-low** method of cost estimation.

b) The **least-square regression** method of cost estimation.

7 Cost-Volume-Profit Analysis

LEARNING OBJECTIVES

- Develop cost-volume-profit models.
- Compute and explain **contribution margin** (CM) and **contribution margin ratio** (CMR).
- Prepare a **contribution-margin-format income statement**.
- Distinguish the contribution-margin-format and the gross-margin format income statements.
- State the underlying assumptions of basic cost-volume-profit (CVP) analysis.
- Compute and explain the margin of safety for a product.
- Compute break-even sales in quantity and revenue for a single product.
- Compute break-even sales in quantities and revenues for multiple products.
- Compute sales volume required to earn the desired amount of profit.
- Understand operating leverage.
- Explain sales mix and the impact of changes in sales mix on contribution margin, break-even point, and profit of a firm.

7.1. INTRODUCTION

Cost-volume-profit (CVP) analysis, also known as **break-even analysis**, is an analytical approach to determine how changes in the volume (quantity) of an activity (e.g., production) affect a company's costs and profit. This analysis is based on the behavior of costs and expenses as variable costs (varying in direct proportion to production volume) or fixed costs (remaining unchanged over a relevant range as defined in the previous chapter). The components of CVP analysis are:

1) Level or volume of activity.
2) Selling prices (per unit).
3) Variable cost per unit (including direct material costs, direct labor costs, and any expenses that vary in proportion to the units produced and sold).
4) Total fixed costs (including the cost of goods manufactured and sold, and all operating period expenses).

DOI: 10.1201/9780429432163-7

7.2. LINEAR COST-VOLUME-PROFIT MODELS FOR A SINGLE PRODUCT

The conventional cost-volume-profit (CVP) analysis is conducted to determine whether a firm is able to achieve a specific level of profit. It is a powerful tool for examining the behavior of total revenues, total costs, and net income as changes occur in activity level, selling price, variable costs per unit, and/or fixed costs. However, there are factors that can overpower the CVP analysis. Examples of such factors include general economic conditions, competition, and government regulations.

The profit (before income taxes) can be determined using the following obvious formula:

$$\textbf{Profit}\left(\textbf{Loss}\right) = \textbf{Sales Revenue} - \textbf{Total Cost} \tag{7.1}$$

Therefore, for a product or service to be profitable, its sales revenue must exceed its costs. As the total cost is the sum of total variable and fixed costs (including the operating and finance expenses), we can restate the profit formula as:

$$\text{Profit} = \text{Sales Revenue} - \text{Variable costs} - \text{Fixed costs}$$

The total costs include the cost of goods sold and operating and financing expenses. The above formula can be expressed in the following mathematical form:

$$Z = PQ - VQ - F \tag{7.2}$$

By combining the sales revenue (PQ) and total variable cost (VQ), we obtain the following formula for computing the before-tax net income (profit):

$$Z = \left(P - V\right)Q - F \tag{7.3}$$

This equation is called the **conventional cost-volume-profit** (CVP) model or function. In the CVP model, the expression ($P - V$) is called the contribution margin per unit when multiplied by the sales quantity (Q units), and the result will be the total contribution margin generated by producing and selling Q units. Therefore, we compute the total contribution margin (CMT) using the following formula:

$$\text{CMT}\left(\text{total}\right) = \left(P - V\right)Q \tag{7.4}$$

where

P = Sales price ($ per unit);
V = Variable cost per unit ($ per unit);
Q = Sales quantity (units);
$PQ = TR$ = Total revenue by selling Q units;
$VQ = TC$ = Total variable cost of producing Q units.

According to Equation 7.1, a firm can encounter one of the following three situations:

1) If the Sales Revenue < Total Expenses, then Profit is negative (loss).
2) If the Sales Revenue = Total Expenses, then Profit is zero (break-even point).
3) If the Sales Revenue > Fixed Expenses, then Profit is positive.

The same three situations can occur by examining Equation (7.3):

1) If $(P - V)Q < F$, then $Z < 0$ (loss).
2) If $(P - V)Q = F$, then $Z = 0$ (break-even point).
3) If $(P - V)Q > F$, then $Z > 0$ (profit).

Dividing the total contribution margin (Equation 7.4) by the sales quantity (Q) gives the contribution margin per unit (CMU), as follows:

$$\text{CMU}\left(\text{per unit}\right) = \frac{(P - V)Q}{Q} = P - V$$

That is,

$$\text{CMU} = P - V \tag{7.5}$$

The contribution margin per unit represents the incremental net income generated for each unit sold after deducting the variable unit cost. Basically, it shows how sales would help a product (or a firm) to cover the fixed costs.

While the CVP analysis cannot really predict how many units of a product will be sold, it is capable of providing information about the financial consequences of different courses of action. One of the most important decisions that need to be made at the beginning of a fiscal period is about the "number of units of a product need to produce and sell in order to break even or to make a certain amount of profit." "Breaking even" simply means covering all variable and fixed costs without making a profit or incurring a loss.

7.2.1. Underlying Assumptions in CVP Analysis

To present the concepts of the cost-volume-profit analysis, we consider the deterministic CVP models with the following underlying assumptions, regardless of the complexity of the situation:

- All expenses can be accurately classified into either fixed or variable.
- Selling price, variable cost (expense) per unit, and total fixed expenses per period are all known and remain constant through the relevant range.
- There will be no changes in the firm's efficiency or productivity.

- Changes in activity level (production/sales quantity) are the only factor that changes revenues and costs (i.e., the volume is the only variable in the CVP model).
- Inventory level does not change from period to period; that is, all units produced are sold. Thus, beginning inventory = ending inventory.
- In multi-product situations, the product mix (their relative sales proportions) will remain constant at all levels of sales.
- The CVP analysis is used on a short-run basis. Therefore, there is no uncertainty, risks, or inflation. However, if they exist and are known, they are incorporated in the CVP model.

The assumptions may seem unrealistic in many situations. Nevertheless, they greatly simplify the CVP analysis. Despite its deficiencies, the CVP analysis provides useful information for performing the break-even analysis and in short-term goal settings (such as sales plans, profit targets, production budgets, and pricing strategies).

7.2.2. Marginal Cost and Revenue and Contribution Margin

Recalling that the total cost (TC) is the sum of total variable costs and total fixed costs, mathematically expressed as:

$$\boldsymbol{TC = V \cdot Q + F} \tag{7.6}$$

where

V = Variable cost per unit;
Q = The production quantity (units produced);
F = The total fixed cost;
$V{\cdot}Q$ = Total variable cost.

The ***marginal cost*** is the additional cost incurred to produce an additional unit of a product, which is the rate of change in the total cost, found by taking the derivative of the total cost function with respect to production quantity (Q), that is,

$$\text{MC} = \frac{dTC}{dQ} = V \tag{7.7}$$

Note that the fixed cost (F) remains unchanged regardless of the changes in the production volume, but total cost increases by V dollars (variable cost per unit) for every additional unit produced. That is, the variable cost per unit (V) is the slope of the cost curve (function).

The **total revenue** (TR) is the sales price multiplied by the quantity of the product sold, mathematically expressed as:

$$\textbf{TR} = \boldsymbol{P \cdot Q} \tag{7.8}$$

The **marginal revenue** (***MR***) is the additional revenue generated by selling an additional unit of the product, which is the rate of change in the total revenue. It is the derivative of the revenue function (Equation 7.8) with respect to sales quantity (Q), that is:

$$MR = P \tag{7.9}$$

Taking the derivative of the CVP model (Equation 7.3) or the derivative of the contribution margin formula (Equation 7.4) with respect to the sales quantity (Q) gives the marginal profit, which is the contribution margin per unit.

$$\text{CMU} = \frac{\text{dZ}}{\text{dQ}} = P - V \tag{7.10}$$

Therefore, the **contribution margin per unit (CMU)** means that for every (additional) unit of the product manufactured and sold, it contributes $P - V$ dollars toward covering the fixed costs and making profit.

The crucial part of CVP analysis is the contribution margin (CM) and the **contribution margin ratio (CMR)**. The contribution margin represents the part of income or profit the company made before deducting its fixed costs. In other words, it is the portion of the sales amount available to cover (or contribute to) fixed costs. The contribution margin ratio (CMR) is a percentage of sales revenue available to cover fixed costs. Once fixed costs are covered, the remainder of sales is the net income (profit). The CMR is computed as follows:

$$\text{CMR} = \frac{\text{CM}}{\text{Sales Amount}} = \frac{(\text{Sales Amount}) - (\text{Total Variable Cost})}{\text{Sales Amount}}$$

$$\text{CMR} = \frac{PQ - VQ}{PQ} = \frac{P - V}{P} \tag{7.11}$$

where

P = Sales price per unit;
V = Variable cost per unit;
Q = the production quantity (units produced);
PQ = Sales value (sales amount in dollars);
VQ = Total variable cost.

Example 7.1

A manufacturing company produces a product that sells for $5.60 per unit, whereas its variable costs and expenses are $1.40 per unit. The company has an annual fixed cost of $504,000. The company's projected sales in the upcoming year are 180,000 units of its product. Compute the company's contribution margin (CM), contribution margin ratio (CMR), and expected profit in the year.

Solution

Contribution margin per unit:

$$CMU = P - V = \$5.60 - \$1.40 = \$4.20 \text{ per unit}$$

Total contribution margin

$$CMT = (P - V)Q = \$4.20(180{,}000) = \$756{,}000$$

Contributing margin ratio using total sales and total contribution margin:

$$CMR = \frac{CM}{Sales} = \frac{(P-V)Q}{PQ} = \frac{(\$5.60 - \$1.40)(180{,}000)}{\$5.60(180{,}000)} = 0.75 = \mathbf{75\%} \text{ of sales}$$

Contributing margin ratio using sales price and contribution margin per unit:

$$CMR = \frac{CM}{Sales} = \frac{P-V}{P} = \frac{\$5.60 - \$1.40}{\$5.60} = 0.75 = \mathbf{75\%} \text{ of sales}$$

The expected profit using the CVP equation:

$$Z = (P - V)Q - F$$

$$Z(180{,}000) = (\$5.60 - \$1.40)(180{,}000) - \$504{,}000 = \mathbf{\$252{,}000}$$

7.2.3. Break-even Analysis for a Single Product

A critical part of CVP analysis is the break-even point, the level of activity (production level), where the total revenues equal total costs (both fixed and variable costs). At the break-even point, a company will experience zero profit (i.e., no profit no loss).

If a company's costs were all variable, the problem of breakeven would never arise. However, as usually some costs are variable and others are fixed, the company, as previously stated, suffers losses up to a certain level of sales (when revenue equals total costs), beyond which the company would have profit. This type of analysis is useful for short-term managerial decision-making.

7.2.3.1. Graphical Method for Finding the Break-even Point

The **break-even point** (**BEP**) is the volume of sales at which total revenue equals total cost. That is, at the break-even point, the profit of the company is zero (no profit, no loss).

The break-even chart is a graphical representation of total revenue and total cost at various levels of sales as well as the variation of income (or sales, revenue) shown on the same chart with the variation in the activity level (*e.g.*, production and sales quantity). Figure 7.1 illustrates the concept of the break-even point.

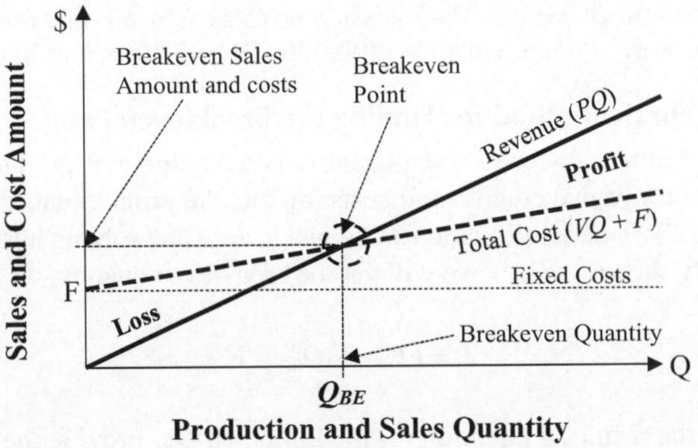

FIGURE 7.1 A graphical illustration of the break-even point.

Example 7.2

A company manufactures a product that sells for $2.00 (per unit) with a variable cost of $1.20 per unit and a total fixed cost of $40,000 per period. Find the break-even point for the company using the graphical method.

Solution

As Figure 7.2 shows, up to the break-even point, found at the intersection of the total-revenue and total-costs curves, the firm suffers losses. Beyond this point, the firm makes profits. The **break-even point** is the cross point of the sales revenue and the total cost lines.

If we use a graph paper, we can find the **break-even quantity** (Q_{BE} = 50,000 units) by drawing a vertical line from the cross points of total revenue and total cost lines to intercept the horizontal axis, which is scaled to represent the production and sales volume (units). The **break-even sales amount** (revenue) is the intercept

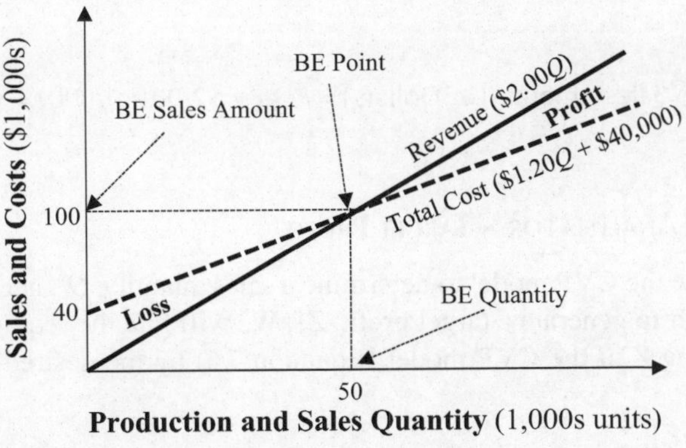

FIGURE 7.2 The break-even point chart for Example 7.2.

point on the vertical axis (i.e., the sales and costs axis) by a horizontal line drawn from the break-even point, which is \$100,000 (TR_{BE} = \$2.00×50,000 units)

7.2.3.2. Algebraic Method for Finding the Break-even Point

As previously defined, the break-even quantity is the volume of production and sales at which the total revenue equals total costs; that is, the profit equals zero. Hence, by setting Z in the CVP model (Equation 7.6) equal to zero and solving it for the unknown Q (symbolically shown as Q_{BE}), we will get the break-even quantity, as follows:

$$0=(P-V)Q_{BE}-F$$

Then, solving the resulting equation for the unknown Q_{BE} provide the following formula for computing the **break-even quantity**:

$$Q_{BE}=\frac{F}{(P-V)} \tag{7.12}$$

Again, the expression in the denominator is the amount contributed by every unit produced and sold towards covering the fixed costs.

Example 7.3

For the company in Example 7.2, find the break-even point using the algebraic method.

Solution

We find the break-even quantity to be:

$$Q_{BE}=\$40{,}000/(\$2.00-\$1.20)=50{,}000 \text{ units}$$

As done in Example 7.2, we can calculate the break-even sales amount (revenue) as follows:

$$\text{Breakeven Sales Amount (or Dollars)}=PQ_{BE}=\$2.00(50{,}000)=\$100{,}000.$$

7.2.4. CVP Analysis for a Target Profit

We can also use the CVP model to determine a sales quantity, Q^*, needed to produce and sell in order to generate a target profit, Z^*. We will find the required sales quantity by replacing Z in the CVP model (Equation 7.6) by the desired profit amount, Z^*, as follows:

$$Z^*=(P-V)\cdot Q^*-F$$

Then, we solve the equation for the unknown sales quantity, Q^*, which not only covers the total fixed cost (F) but also provides the target profit, Z^*, as expressed by the following formula:

$$Q^* = \frac{Z^* + F}{P - V} \tag{7.13}$$

Example 7.4

For the scenario given in **Examples 7.2 and 7.3**, determined the required sales revenue that provides a desired (target) profit of \$20,000.

Solution

We can directly use Equation (7.15) to compute the desired sales quantity as:

$$Q^* = \frac{Z^* + F}{P - V} = \frac{\$20{,}000 + \$40{,}000}{\$2.00 - \$1.20} = \mathbf{75{,}000\ units}$$

Alternatively, we can set $Z = Z^* = \$20{,}000$ in the CVP model (Equation 7.3) and compute the required sales quantity as follows:

$$Z^* = (P - V) \cdot Q^* - F$$

$$\$20{,}000 = (\$2.00 - \$1.20)Q^* - \$40{,}000$$

$$Q^* = \$60{,}000/\$0.80 = 75{,}000 \text{ units}$$

Then, we compute the required sales revenue as follows:

$$TR^* = P \cdot Q^* = \$2(75{,}000) = \$150{,}000$$

7.2.5. Margin of Safety

The margin of safety (MOS) is the difference between the estimated (or actual) sales and the level by which the sales amount can decrease before a loss occurs. That is, it is the excess of the estimated (or actual) revenues over the break-even revenues as represented by the following formula:

$$\textbf{MOS} = \textbf{Actual (or Estimated) Sales Amount} - \textbf{Breakeven Sales} \tag{7.14}$$

The margin of safety is usually expressed as a percentage of the estimated sales:

$$\textbf{MOS\%} = \frac{\textbf{MOS}}{\textbf{Actual (or Estimated) Sales}} \tag{7.15}$$

MOS is an indicator of the risk of loss when there is a possibility of a significant sales decline. A business with a low margin of safety percentage may attempt expenses cutting, while a high spread of margin assures a company that it is protected against sales variability. MOS may be expressed by the sales quantity (number of units):

$$\textbf{MOS in Units} = \textbf{Actual}\left(\textbf{or Estimated}\right)\textbf{ Sales Units} - \textbf{Breakeven Quantity} \tag{7.16}$$

Example 7.5

A company expects the sales price for its product in the upcoming year to be about \$70 and its variable costs are approximately \$30. The company has estimated annual fixed costs of \$250,000. If the company's projected sales are 10,000 units of the product, what will be its margin of safety?

Solution

Contribution margin per unit: CMU = \$70 – \$30 = \$40

The break-even quantity: $Q_{BE} = \dfrac{\text{Fixed Cost}}{CM_U} = \dfrac{\$250{,}000}{\$40} = 6{,}250$ units

The break-even sales amount = \$70(6,250) = \$437,500

The projected sales amount = \$70(10,000) = \$700,000

The Margin of Safety: MOS = Projected sales amount – break-even sales amount

= \$700,000 – \$437,500 = \$262,500

The Margin of Safety percentage: MOS % = \$262,500/\$700,000 = 0.375 = 37.5%

7.2.6. Operating Leverage

The operating leverage measures the degree to which a firm or product can increase its profit by increasing revenue. If a firm's fixed cost is higher in proportion than its variable cost, it would have high operating leverage and will generate a larger profit from an incremental sale.

The operating leverage is calculated by dividing the contribution margin by the difference between the operating income.

$$\textbf{Operating Leverage} = \frac{\textbf{Contribution Margin}}{\textbf{Operating Income}} = \frac{(P-V)Q}{(P-V)Q-F} \tag{7.17}$$

A firm (or product) with low operating leverage may have high variable costs that vary directly with the sales volume but has lower fixed costs to cover each period.

Example 7.6

A company sells 45,000 units of its product at \$10 each. The company's costs are \$90,000 fixed costs and \$2 variable costs per unit. Determine the company's degree of operating leverage.

Solution

$$\text{Operating Leverage} = \frac{\text{Contribution Margin}}{\text{Profit}}$$

$$= \frac{(\$10-\$2)(45{,}000)}{(\$10-\$2)(45{,}000)-\$90{,}000} = 1.33 \text{ or } 133\%$$

An operating leverage of 133% means a 10% sales increase would result in a 13.3% increase in operating income, which is $1.33 \times 10\%$.

7.3. NONLINEAR CVP ANALYSIS FOR A SINGLE PRODUCT

In previous CVP analysis, the assumptions were that the sales price (P), fixed cost (F), and unit variable cost per unit (V) are all constant at any production level. These assumptions lead to the linearity of the model developed for CVP analysis. However, if the sales price varies as a function of quantity supplied to the market; that is, $P=f(Q)$, which corresponds to the inverse function of the economists' demand curve, $Q=f^{-1}(P)$. This function presents the sales quantity as a function of the price, which decreases as the price increases.

The total cost of production, TC(Q), could also be a nonlinear function of the production level, in which not only the marginal cost varies but also the fixed cost can shift to higher levels, beyond certain production levels. In general, the equation for CVP is written as

$$Z(Q) = Q \cdot f^{-1}(P) - TC(Q)$$

Let's assume that the total cost function takes the form of the following polynomial.

$$TC(Q) = a_0 + a_1Q + a_2Q^2 + a_3Q^3,$$

where a_0, a_1, a_2, and a_3 are constant coefficients. Then, the marginal-cost function, $MC(Q)$, is obtained by taking the derivative of the total-cost function:

$$MC(Q) = \frac{dTC(Q)}{dQ} = a_1 + 2a_2Q + 3a_3Q^2$$

Figure 7.3 illustrates how a total cost of a product may fluctuate as the production quantity changes when the cost is a non-linear function of production level.

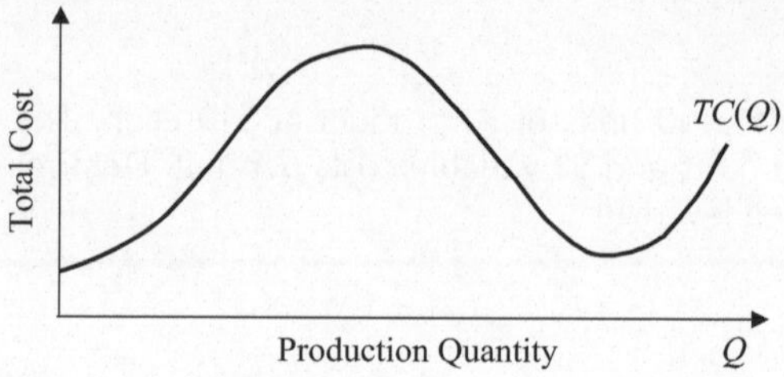

FIGURE 7.3 An example of a nonlinear cost curve.

Assuming that the demand for the products is a linear function of price, that is,

$$Q = b_0 + b_1 \cdot P,$$

where b_0 is a constant and $b_1 < 0$ is the slope (also constant). The negativity of b_1 is the indication of that sales decreases as a reaction to the increases in sales price. The total revenue function, TR(Q), is then given by the following quadratic function.

$$\text{TR}(Q) = Q \cdot P = Q(Q - b_0)/b_1$$

Figure 7.4 illustrates non-linear changes in total-revenue and the total-cost functions as the level of activity/production changes.

In this example of nonlinear revenue and cost, we see that there are two break-even points associated with the production levels, where the total-revenue and total-cost curves intersect; that is, the total cost equals total revenue. Therefore, we can find the break-even point quantities by solving the following nonlinear equation.

$$\text{TR}(Q) = \text{TC}(Q)$$

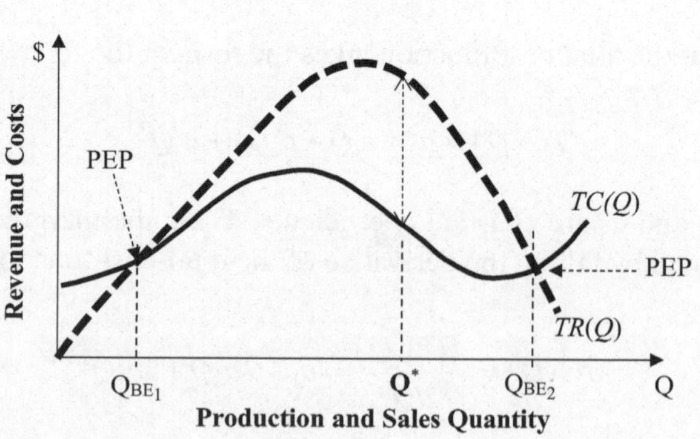

FIGURE 7.4 An illustration of nonlinear revenue and cost functions.

Substituting TR(Q) and TC(Q) by their corresponding expressions give the following nonlinear equation:

$$\left(Q^2 - mQ\right)/n = a + bQ + cQ^2 + dQ^3$$

Solving this equation for the unknown Q gives the break-even quantity(ies). However, we may be more interested in finding a production level that maximizes the profited.

For such a problem, there are two general methods for finding the optimal production output, which maximizes the firm's profits. We explain these methods as follows:

- **Economists' Method**: According to the economics principles, when the marginal revenue equals marginal cost, profit will be at its maximum; that is,

  $$MR(Q) = MC(Q)$$

  or

  $$b + 2cQ + 3dQ^2 = (2Q - m)/n$$

  when solved for Q, the production level (Q^*) which provides the maximum profit (Z^*) is found.
- **Mathematical Method**: By subtracting the cost function from the revenue function, the profit function, $Z(Q)$, or the CVP model, is obtained. Then, by setting the first derivative of $Z(Q)$ equal to zero, we find a production level (Q^*) that yields the maximum profit (Z^*).

  Note that the optimal value of Q^* is the number of units that have to be produced and sold to maximize profits. Substituting Q by Q^* in the total revenue and total cost functions, we find their optimal amount. The difference between these two values is the maximum profit. We can also be substitute Q^* for Q in the profit function to find the maximum amount of profits.

Example 7.7

Assume that the total cost, TC(Q), and total-revenue, TR(Q), respectively, are given as:

$$TC(Q) = 504 + 20Q - 3Q^2 + 0.5Q^3$$

and

$$TR(Q) = 564Q - 8Q^2$$

Then the CVP model is formulated as follows:

$$Z = TR(Q) - TC(Q)$$

$$= 564Q - 8Q^2 - \left(504 + 20Q - 3Q^2 + 0.5Q^3\right)$$

That is,

$$\mathbf{Z = -0.5Q^3 - 5Q^2 + 544Q - 504} \qquad \text{(CVP model)}$$

To find the break-even points, we set TR(Q) equal to TC(Q) (or Z=0). Either method results in the following equation:

$$-0.5Q^3 - 5Q^2 + 544Q - 504 = 0$$

Solving this equation graphically (Figure 7.5) or by trial and error yields the two break-even points

$$Q_{BE} = 0.94 \quad \text{and} \quad Q_{BE} = 27.81.$$

Therefore, the product will be profitable as long as the level of production is in the following range:

$$0.94 < Q < 27.81.$$

The maximum profit is achieved when marginal revenue, MR(Q), equals marginal cost, MC(Q).

$$MR(Q) = 564 - 16Q$$

$$MC(Q) = 20 - 6Q + 1.5Q^2$$

To maximize the profit, set $MR(Q) = MC(Q)$

$$564 - 16Q = 20 - 6Q + 1.5Q^2$$

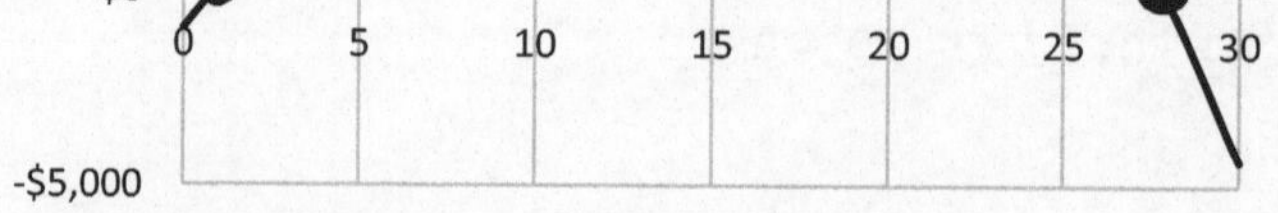

FIGURE 7.5 Graphical method for finding the break-even points (Example 7.7).

We can rearrange the equation and write it as

$$1.5Q^2 + 10Q - 544 = 0$$

Then, solving the quadratic equation for Q (where Q>0), gives

$$Q^* = \frac{-10 \pm \sqrt{10^2 - 4(1.5)(-544)}}{2(1.5)} = 16 \text{ units}$$

At this point, both marginal revenue and marginal cost are equal to $308 as computed below.

$$MR(16) = 564 - 16(16) = \$308$$

$$MC(16) = 20 - 6(16) + 1.5(16)^2 = \$308$$

Then, the maximum profit is computed as follows:

$$TR(Q) = 564Q - 8Q^2$$

$$TR(16) = 564(16) - 8(16)^2 = \$6,976$$

and

$$TC(Q) = 504 + 20Q - 3Q^2 + 0.5Q^3$$

$$TC(16) = 504 + 20(16) - 3(16)^2 + 0.5(16)^3 = \$2,104$$

Then, profit is found to be:

$$Z^* = TR(16) - TC(16) = \$6,976 - \$2,104 = \mathbf{\$4,872}$$

Alternatively, we could maximize profit by setting the first derivative of the CVP equation equal to zero, as follows:

$$Z = -\,0.5Q^3 - 5Q^2 + 544Q - 504 \quad \text{(CVP model)}$$

$$\frac{dZ}{dQ} = -1.5Q^2 - 10Q + 544 = 0$$

$$Q^* = \frac{10 \pm \sqrt{10^2 - 4(-1.5)(544)}}{2(-1.5)} = \frac{10 \pm \sqrt{10^2 - 4(-1.5)(544)}}{2(-1.5)} = \frac{10 \pm 58}{-3}$$

$$Q^* = -22.67 < 0 \text{ (not acceptable)}$$

$$\mathbf{Q^* = 16}$$

Then, the maximum profit, Z*, is computed using the CVP equation as follows:

$$Z^* = Z(Q^*) = Z(16) = -0.5(16)^3 - 5(16)^2 + 544(16) - 504$$

$$\mathbf{Z^* = \$4{,}872}$$

7.4. COST-VOLUME-PROFIT ANALYSIS FOR MULTIPLE PRODUCTS

In this section, we present the procedures for performing the CVP analysis for multi-product situations. Multi-product CVP analysis arises when a company sells two or more products. The procedure of computing the break-even points or required sales volumes to achieve a target profit in a multiple-product problem can be more complicated than that is used in single-product cases.

A general CVP model for multiple (say n) products can be written as:

$$Z = \sum_{i=1}^{n} (P_i - V_i) Q_i - \sum_{i=1}^{n} F_i \tag{7.18}$$

subject to the non-negativity constraints on the production volumes, that is,

$$Q_i \geq 0,\ i = 1,\ 2, \ldots, n.$$

where

P_i = Sales price (\$/unit) for product i (for $i = 1, 2, \ldots, n$);
V_i = Variable cost per unit of product i (\$/unit) (for $i = 1, 2, \ldots, n$);
F_i = Fixed cost assigned to product i (\$/period) (for $i = 1, 2, \ldots, n$);
Q_i = Production volume (units) for product i (for $i = 1, 2, \ldots, n$);
$(P_i - V_i)$ = Contribution margin per unit for product i (for $i = 1, 2, \ldots, n$);
$(P_i - V_i)Q_i$ = Total contribution margin for product i (for $i = 1, 2, \ldots, n$);
Z = Profit (loss).

7.4.1. CVP Analysis for Multiple Products with a Fixed Sales Mix

When the projections of sales are based on the assumption that the same sales mix of products and contribution margin ratios, the CVP analysis can be performed using either of the following two methods:

7.4.1.1. Product-Indexing Method for Sales-Mix CVP Analysis

The analyst can use this method following the steps below:

Step 1.1. Choose one of the products as an index product (named as product i) and express the sales quantity (Q_j) of each of other products in terms (i.e., a multiple) of the sales quantity of the index product based on the sales-mix quota, as $Q_j = K_j \cdot Q_i$.

Step 1.2. Replace each product's quantity (Q_j) in the multiple-product CVP model with its equivalent in terms of the index product's quantity ($K_j \cdot Q_i$). This converts the CVP model into a single-product model.

Step 1.3. Compute the required sales volume (Q_i) for the index product to break even or to achieve a target profit. Then, compute the required sales volumes for other products by multiplying the index product's sales volume by their corresponding multiplier, K_j, determined in **Step 1.1**, as $Q_j = K_j \cdot Q_i$.

7.4.1.2. Weighted-Average Contribution Margin Method for Sales-Mix CVP Analysis

The analyst can use this method following the steps below:

Step 2.1. Compute the total contribution margin for each product and sum the totals to determine the **combined total contribution margin** for all products.

Step 2.2. Add the sales quantities (units) for all products into the **combined total sales units.**

Step 2.3. Divide the combined total contribution margin (from **Step 1**) by the combined total sales units (from **Step 2**) to determine the weighted-average contribution margin (WACM) per composite unit, as follows:

$$\text{WACM} = \frac{\text{Combined Contribution Margin}}{\text{Combined Sales Units}}$$

Step 2.4. Compute the composite sales volume (Q) required to break even or to achieve a target profit, as follows:

Break-even Composite Units:

$$Q_{\text{BE}} = \frac{\textbf{Total Fixed Costs}}{\textbf{WACM Per Composite Unit}}$$

Required Composite Units to Achieve Target Profit:

$$Q^* = \frac{(\textbf{Target Profit}) + (\textbf{Total Fixed Costs})}{\textbf{WACM Per Composite Unit}}$$

Step 2.5. Then, compute the required sales volumes for the individual products using their corresponding sales-mix proportion of the overall composite units.

Example 7.8

A company manufactures and sells three products, A, B, and C. The weekly sales and variable costs of the products are shown below:

Product	Sales Units	Sales Price	Total Sales	Variable Cost per Unit
A	25,000	$50	$1,000,000	$20
B	10,000	100	1,000,000	50
C	5,000	200	1,000,000	120
Totals	**40,000**		**$3,000,000**	

If the company's fixed costs are $495,000 per year and maintain the product-mix ratios constant, calculate the break-even points for the products using:

a) Product-indexing method.
b) Weighted-average contribution-margin method.

Solution

a) Finding the break-even quantities using the product-indexing method.
The product-mix quotas are 25,000:10,000:5,000 or 5:2:1. Thus, we should choose product C as the index product use $Q_A = 5Q_C$ and $Q_B = 2Q_C$. in the CVP formulation, as follows:

$$Z = (P_A - V_A)Q_A + (P_B - V_B)Q_B + (P_C - V_C)Q_C - 495{,}000$$

$$Z = (50 - 20)(5Q_C) + (100 - 50)(2Q_C) + (200 - 120)Q_C - 495{,}000$$

$$Z = 150Q_C + 100Q_C + 80Q_C - 495{,}000$$

$$Z = 330Q_C - 495{,}000$$

To break even, set $Z = 0$ and solve for $(Q_C)_{BE}$:

$$0 = 330(Q_C)_{BE} - 495{,}000$$

$$\mathbf{(Q_C)_{BE} = 1{,}500 \text{ units}}$$

Then, we compute the break-even quantities for products A and B as follows

$$\mathbf{(Q_A)_{BE} = 5(1{,}500) = 7{,}500 \text{ units}}$$

$$\mathbf{(Q_B)_{BE} = 2(1{,}500) = 3{,}000 \text{ units}}$$

b) Finding the break-even quantities using the weighted-average contribution-margin method.

The computations are presented in the following table:

Product	Sales Units	Sales Price	Total Sales	Variable Cost per Unit	Contribution Margin/Unit	Contribution Margin
A	25,000	$ 50	$1,250,000	$ 20	$30	$750,000
B	10,000	100	1,000,000	50	50	500,000
C	5,000	200	1,000,000	120	80	400,000
Totals	40,000		$3,250,000			$1,650,000

$$\text{WACM Ratio} = \frac{\text{Combibed Contribution Margi}}{\text{Composite Total Sales}} = \frac{\$1,650,000}{\$3,250,000} = 50.77\%$$

$$\text{WACM} = \frac{\text{Combibed Contribution Margin}}{\text{Combined Sales Units}} = \frac{\$1,650,000}{40,000 \text{ Units}} = \$41.25 \text{ per unit}$$

Using the weighted-average contribution margin (WACM) we can calculate the composite break-even point quantity as follows:

$$\text{Composite Breakeven Point Quantity} = \frac{\text{Fixed Costs}}{\text{WACM}} = \frac{\$495,000}{\$41.25/\text{unit}} = 12,000 \text{ units}$$

We can compute the break-even sales quantity (in units) required for each product based on the proportion of each product in the sales mix:

$$(Q_A)_{BE} = \frac{25,000}{40,000}(12,000) = \mathbf{7,500 \text{ units}}$$

$$(Q_B)_{BE} = \frac{10,000}{40,000}(12,000) = \mathbf{3,000} \text{ units}$$

$$(Q_C)_{BE} = \frac{5,000}{40,000}(12,000) = \mathbf{1,500} \text{ units}$$

The break-even sales amount (in dollars) for each product depends on the proportion of each product in the sales mix:

$$\text{Breakeven sales value for Product A} = \$50(7,500) = \$375,000$$

$$\text{Breakeven sales value for Product B} = \$100(3,000) = \$300,000$$

$$\text{Breakeven sales value for Product C} = \$200(1,500) = \$300,000$$

7.4.2. CVP Analysis for Multiple Products without Product Interdependencies

In general, in the absence of the interdependency among the products, to determine a product mix that yields a given targeted profit level, Z^* including the break-even points ($Z^* = 0$), we will seek the feasible solutions to the following CVP Model:

$$Z^* = \sum_{j=1}^{n} (P_j - V_j) Q_j - \sum_{j=1}^{n} F_j \tag{7.19}$$

subject to the non-negativity constraints on the production volumes, that is,

$$Q_j \geq 0, \ \text{for } j = 1, 2, \ldots, n$$

However, there would be an infinite number of solutions to Equation (7.19), which at the same time satisfy the non-negativity constraints. For instance, one solution is:

$$Q_1^* = \frac{Z^* + \sum_{j=1}^{n} F_j}{P_1 - V_1}, Q_2^* = Q_3^* = \ldots = Q_n^* = 0$$

and another solution is

$$Q_1^* = 0, Q_2^* = \frac{Z^* + \sum_{j=1}^{n} F_j}{P_2 - V_2}, Q_3^* = \ldots = Q_n^* = 0$$

We can find a total of n different solutions in this same fashion. In addition to these n solutions, any linear combinations of these n solutions would also be another feasible solution. Therefore, there are an infinite number of solutions that would also yield the targeted profit level. For example, if the above two basic solutions are shown by $Y^{(1)}$ and $Y^{(2)}$, that is,

$$Y^{(1)} = \text{the first solution vector } (Q_1, Q_2, Q_3, \ldots, Q_n) \text{ and}$$

$$Y^{(2)} = \text{the second solution vector } (Q_1, Q_2, Q_3, \ldots, Q_n),$$

then

$$mY^{(1)} + (1 - m) Y^{(2)} \quad \text{where } 0 \leq m \leq 1$$

would be a solution to Equation (7.18).

This means that any linear combination of the n basic solutions $\sum_{i=1}^{n} m_i Y^{(i)}$ (where $m_i \geq 0$ and $\sum_{i=1}^{n} m_i = 1$) would also be a solution to Equation (7.18).

Example 7.9

Consider a two-product situation with the following data:

Product	P_i	V_i	F_i
1	$22	$6	$20,000
2	30	10	30,000

The break-even point for each product is given below:

$$\text{Product 1}: Q_{1,BE} = 20{,}000/(22-6) = 1{,}250 \text{ units}$$

$$\text{Product 2}: Q_{2,BE} = 30{,}000/(30-10) = 1{,}500 \text{ units}$$

The profit equation for this problem is found as follows:

$$Z = \left[(22-6)Q_1 - 20{,}000\right] + \left[(30-10)Q_2 - 30{,}000\right]$$

$$Z = 16Q_1 - 20{,}000 + 20Q_2 - 30{,}000$$

$$Z = 16Q_1 + 20Q_2 - 50{,}000$$

In order to break even, the sales quantity vector (Q_1, Q_2) must be selected such that

$$16Q_1 + 20Q_2 - 50{,}000 = 0$$

or $$16Q_1 + 20Q_2 = 50{,}000$$

Two of the feasible solutions are:

$$Y^{(1)} : Q_1 = 3{,}125,\ Q_2 = 0$$

and

$$Y^{(2)} : Q_1 = 0,\ Q_2 = 2{,}500$$

Also, any product mixes of the form $m \cdot Y^{(1)} + (1 - m) \cdot Y^{(2)}$, where $0 \le m \le 1$, will break even (i.e., $Z = 0$). For example, if we choose $m = 0.4$, then another solution is:

$$0.4Y^{(1)} + (1-0.4)Y^{(2)} = 0.4(3{,}125, 0) + 0.6(0, 2{,}500) = (1{,}250, 1{,}500).$$

That is, we can break even with $Q_1 = 1{,}250$ and $Q_2 = 1{,}500$. An infinite number of product mixes can be found by letting m vary continuously in the range from 0 to 1, which yields zero profits (i.e., breaking even).

For a given targeted profit, Z^*, any product mix (Q_1, Q_2) that satisfies

$$16Q_1 + 20Q_2 = 50,000 + Z^*$$

would be a solution and provide an overall profit level of Z^*.

For problems of this type, when there are no production requirements or constraints on the resources by which the capacity of production will be limited, there will be an infinite number of solutions yielding a given targeted profit level including the break-even point. In the next sections, the impacts of production constraints are studied.

7.4.3. CVP Analysis for Multiple Products with a Single Constraint

Achieving the maximum profit is usually the most important factor in the mind of a cost analyst. In the CVP analysis, products with higher contribution margins ($Pi - Vi$) are given special attention. The reason for this is that the products with high contribution margins maximize the overall contribution margin and, automatically, profit. But, usually, there are constraints on the company's resources that limit the production output.

To make a meaningful CVP analysis, one should identify all production constraints. Examples of such constraints are the limitations on the availability of machine time, labor, supervision, materials, storage space, and transportation, as well as production requirements and the existence of interdependencies among products.

In this section, we concentrate on the single constraint situation. In such a situation, the objective is to maximize the production of products that yield the highest contribution margin per unit of the limited resource.

Example 7.10

In the previous example, assume that only 30,000 hours of machine time are available. Assume, also, that each unit of Product 1 requires 3 machine hours and each unit of Product 2 requires 4 machine hours. The situation is summarized in the following table.

Product	Contribution Margin Per Unit	Machine Time Required Per Unit
1	$16	3 hours
2	20	4 hours

When considering the contribution margins, per unit of the individual products, Product 2, with $4 per unit more contribution margin seems to be more profitable than Product 1. This would have been true if either there were no constraints on the production outputs due to the number of machine hours available to produce the two products. However, if the company produces and sells only Product 2, its maximum production output will be 7,500 units of Product 2 (i.e., 30,000 hours/4 hours per unit), which yields a total contribution margin of $150,000 (i.e., 7,500 units × $20). Whereas, if the company only produces and sells Product 1, it can produce 10,000 units (i.e., 30,000 hours/3 hours per unit), which generate a total contribution margin of \$160,000 (i.e., 10,000 units × \$16). Therefore, Product 1 is more profitable than Product 2, although it has a lower contribution margin per unit than does Product 2.

Thus, when there are some constraints on the production resources, ranking the profitability of the products based on contribution margin as the only criteria is misleading. The **contribution margin per unit of the limited resource** provides a suitable criterion for ranking the products in terms of their profitability. The contribution margin of the products per machine hour is found to be:

For Product 1: \$16/3 hours = \$5.33 per machine hour;
For Product 2: \$20/4 hours = \$5.00 per machine hour.

It became obvious that the machine hours would be more profitably used for manufacturing Product 1.

7.4.4. CVP Analysis for Multiple Products with Multiple Constraints and Unavoidable Fixed Costs

Business problems in the real world are not always simple. The resources are usually scarce. In production situations, there are always shortages in capital, time, materials, labor, demand for products, etc. In this section, we focus on the situations in which there are more than one production constraints.

Assume that a company manufactures and sells n different products while it has m constraining resources. Let

i = Product index ($i = 1, 2, \ldots, n$);
j = The jth constraining resource ($j = 1, 2, \ldots, m$);
R_j = Maximum amount of resource j available;
r_{ij} = The fixed amount of resource j required to produce one unit of Product i;
Q_i = The amount of product i produced.

Therefore, to produce Q_1, Q_2, …, and Q_n units of products 1, 2, …, and n, respectively, the following total amount of resource j is required:

$$r_{1j}Q_1 + r_{2j}Q_2 + \ldots + r_{nj}Q_n, \quad (\text{where } j = 1,2,\ldots,m.)$$

However, there are only R_j units of each resource j available, hence, the production outputs must satisfy the following constraints:

$$\sum_{i=1}^{n} r_{ij}Q_i \le R_j \text{ for all } j = 1,2,\ldots,m; \ (\text{where } i = 1,2,\ldots,n)$$

In order to maximize the overall profit of the company, the total contribution margin is generated by all products. To do so, the problem must be formulated as the following linear programming from:

$$\text{Maximize: } Z = \sum_{i=1}^{n}(P_i - V_i)Q_i - \sum_{i=1}^{n} F_i \quad (\text{where } i = 1,2,\ldots,n)$$

$$\text{Subject to: } \sum_{i=1}^{n} r_{ij}Q_i \le R_j \text{ for all } j = 1,2,\ldots,m; \ (\text{where } i = 1,2,\ldots,n)$$

$$Q_i \ge 0 \quad (\text{where } i = 1,2,\ldots,n)$$

In order to have the standard form of a linear programming problem, we eliminate the constant term $\sum_{i=1}^{n} F_i$ in the objective function. If we assume that all fixed costs are unavoidable in case of a product-line shutdown, then maximizing the total contribution margin will also maximize the overall profit. Under this simplifying assumption, the problem becomes as follows:

$$\text{Maximize: } Y = \sum_{i=1}^{n}(P_i - V_i)Q_i$$

$$\text{Subject to: } \sum_{i=1}^{n} r_{ij}Q_i \le R_j \text{ for all } j = 1,2,\ldots,m; \ (\text{where } i = 1,2,\ldots,n)$$

$$Q_i \ge 0 \quad (\text{where } i = 1,2,\ldots,n)$$

Example 7.11

For the previous example, assume that a maximum of 24,000 gallons of raw material is available. Furthermore, the material required to produce one unit of either product is 3 gallons. The following table summarizes the information given for this example.

Product(i)	**Contribution margin per unit ($P_i - V_i$) (\$)**	**Fixed cost (F_i) (\$)**	**Machine time required per unit (r_{i1}) (hours)**	**Raw material per unit (r_{i2}) (units)**
1	16	20,000	3	3
2	20	30,000	4	3

To achieve the optimal solution (product mix), we solve the following linear programming problem:

$$\text{Maximize: } Y = 16Q_1 + 20Q_2$$

$$\text{Subject to: } 3Q_1 + 4Q_2 \le 30{,}000 \quad \text{(machine time constraint)}$$

$$3Q_1 + 3Q_2 \le 24{,}000 \quad \text{(material constraint)}$$

$$Q_1, Q_2 \ge 0 \quad \text{(non-negativity constraints)}$$

The optimal product-mix solution to this linear programming problem is:

$Q_1 = 2{,}000$ units of product 1, and

$Q_2 = 6{,}000$ units of product 2.

This product mix will generate a total contribution margin of \$152,000, which results in an overall profit of

$$Z = 152{,}000 - (20{,}000 + 30{,}000) = \$102{,}000$$

Since there are only two variables (product volumes) in this problem, this optimal product-mix solution can be obtained using the graphical method of linear programming (Figure 7.6).

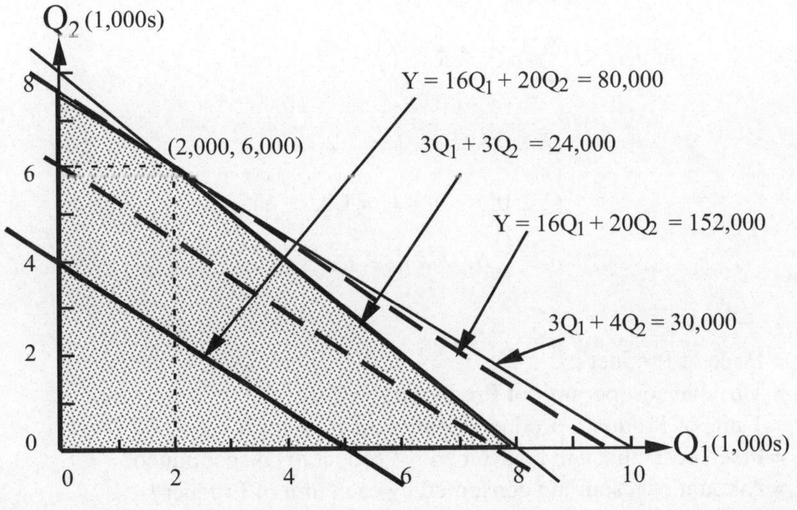

FIGURE 7.6 A graphical solution to the linear program problem, where $Q_1 = 2{,}000$ and $Q_2 = 6{,}000$.

7.4.5. CVP Analysis for Multiple Products with Multiple Constraints and Avoidable Fixed Costs

In the previous sections, we ignored the fixed costs because the maximum value of the total contribution margin guarantees the maximum profits. This was true only if the fixed costs of a product(s) do not disappear by shutting down the production of the corresponding product(s).

There are situations in which if the production of a product is discontinued, some part of its fixed costs will diminish, too. This implies that some part of the total fixed costs is avoidable (or escapable). Here, the linear programming may provide an improper solution to the problem. For example, the linear programming solution may include a product mix with a small production volume of a product whose total contribution margin will not be enough to cover its own avoidable fixed costs, which can be saved if that product line is entirely eliminated.

Examples of avoidable fixed costs are supervisory compensations, cost of leasing equipment, costs of product distribution, and advertising and promotional expenses associated with a product. When there exist fixed costs that are traceable to a specific product line and are avoidable in case of the product-line shutdown, they can be incorporated into the analysis using a more general model. One can determine the optimal product mix by formulating and solving the following form of linear programming model:

$$\text{Maximize: } Z = \sum_{i=1}^{n}(P_i - V_i)Q_i - \sum_{i=1}^{n} q_i F_i \quad (i = 1, 2, \ldots, n)$$

$$\text{Subject to: } \sum_{i=1}^{n} r_{ij} Q_i \le R_j \quad (j = 1, 2, \ldots, m;\ i = 1, 2, \ldots, n)$$

$$Q_i \le q_i \cdot M; \quad (i = 1, 2, \ldots, n)$$

$$Q_i \ge 0; \quad (i = 1, 2, \ldots, n)$$

$$q_i = 0,\ 1 \quad (i = 1, 2, \ldots, n)$$

Where

P_i = Price of Product *i*;
V_i = Variable cost per unit of Product *i*;
Q_i = Units of Product *i* produced (output of Product *i*);
F_i = Fixed costs that can be avoided if Product *i* is discontinued;
r_{ij} = Amount of resource *j* consumed by each unit of Product *i*;
R_i = Maximum amount of resource *j* available;
M = An arbitrary constant that is large enough to be greater than the maximum sales volume of any of the *n* products;
q_i = A binary variable associated with Product *i* so that if $Q_i = 0$, then $q_i = 0$, if $Q_i > 0$, then $q_i = 1$ (the value of q_i indicates whether or not Product *i* is produced.)

The constant M, in the second set of constraints, is an arbitrary large number, in excess of the maximum sales of any product. The constraint $Q_i \leq q_i M$ assures that whenever $Q_i \geq 0$ in a solution, then the variable q_i must be equal to 1, and the escapable fixed cost, F_i, associated with the positive production of Product i will be subtracted from the objective function.

If in a solution $Q_i = 0$, then q_i must also be equal to zero (i.e., $q_i = 0$), and the corresponding avoidable fixed cost will not be subtracted from the objective function.

The traditional linear programming techniques cannot be applied to the fixed-cost problems since they cannot force a decision variable (e.g., q_i) to take on integer values only (i.e., 0 or 1 in this case).

The fixed-cost problem is a special case of a class of integer programming models called mixed-integer programming. The problems are called so because they have both continuous variables (Q_i) and integer (or discrete) variables (q_i).

Example 7.12

The ABC Company produces three products (A, B, and C). Its product-line income statements on a full-cost basis and its projections for the next period are given below. Sales commissions are paid at the rate of 3% of sales to the company's salespersons. Order-processing costs can be considered variable.

Based on the ABC Company's full-cost system, Products B and C are unprofitable, suggesting that overall profits could be increased by dropping both products. However, this computation avoids a careful analysis of the variable contribution margin and escapable fixed costs associated with each product.

	Product			
	A	**B**	**C**	**Total**
Sales volume (Units)	400	300	500	--
Sales revenue ($)	**4,000**	**3,000**	**2,000**	**9,000**
Cost of goods sold:				
Direct material	$1,200	$900	$1,000	$3,100
Direct labor	1,000	1,000	400	2,400
Factory overhead absorbed	400	400	200	1,000
Total cost of goods sold	**$2,600**	**$2,300**	**$1,600**	**$6,500**
Gross margin	**$1,400**	**$700**	**$400**	**$2,500**
Selling expenses:				
Commissions	$ 120	$ 90	$ 60	$ 270
Traceable fixed expenses	200	60	60	320
Common fixed expenses	480	360	240	1,080
Administrative expenses:				
Order processing	30	30	20	80
General administration	320	240	160	720
Total operating expense	**$1,150**	**$780**	**$540**	**$2,470**
Net income (loss) before tax	**$250**	**($80)**	**($140)**	**$30**

Further examination yields the following analysis of factory overhead and fixed operating expenses:

	Product		
	A	B	C
Factory overhead:	$150	$200	$160
Variable (with direct labor cost)			
Fixed but escapable if shut down	150	100	20
Fixed and inescapable if shut down	100	100	20
Traceable fixed selling expense:	180	60	60
Escapable if shut down	20	0	0
Inescapable			
Common fixed selling expense:	200	210	200
Escapable if shut down	280	150	40
Inescapable			
General administrative expense:	20	40	100
Fixed but escapable if shut down	300	200	60
Fixed and inescapable			

The selling prices, *Pi*, for the three products are $10, $10, and $4, respectively. The variable costs and expenses for the three products are:

	Product		
	A	B	C
Variable Costs and Expenses:			
Material	$1,200	$900	$1,000
Labor	1,000	1,000	400
Factory overhead	150	200	160
Sales commissions	120	90	60
Order processing	30	30	20
Total variable costs and expenses	**$2,500**	**$2,220**	**$1,640**
Variable cost per unit	**$6.25**	**$7.40**	**$3.28**
Contribution margin per unit	**$3.75**	**$ 2.60**	**$0.72**

The fixed costs associated with the three products are:

	Product		
	A	B	C
Factory overhead	$250	$200	$ 0
Selling expenses	680	420	300
General administrative	320	240	160
Total fixed costs	**$1,250**	**$860**	**$500**

Our multi-product CVP equation is:

$$Z = 3.75Q_A + 2.60Q_B + 0.72Q_C - (1,250 + 860 + 500)$$

With the projected sales of $Q_A = 400$, $Q_B = 300$, $Q_C = 500$, we obtain:

$$Z = 1,500 + 780 + 360 - 2,610 = \$30$$

which agrees with the profit from the product-line income statement above. With this traditional CVP analysis, it appears that all three products are generating positive contribution margins for the company and hence should be retained if sales will be as projected.

Tracing all fixed costs to the individual products, we can compute the break-even volumes as the ratio of fixed costs to per-unit contribution margin:

$$Q_{A,BE} = 1,250/3.75 = 333 \text{ units}$$
$$Q_{B,BE} = 860/2.60 = 331 \text{ units}$$
$$Q_{C,BE} = 500/0.72 = 694 \text{ units}$$

We see that the projected sales for both products B and C are below the break-even volume, which explains why both products show losses on the product-line income statement. To achieve breakeven, product B must have 10% more sales, while product C must increase sales by almost 40%.

This analysis, however, is misleading, since a substantial part of the fixed costs associated with the products would not disappear were these products to be dropped from the firm's product line. A more meaningful product-line profitability analysis would use only the escapable fixed costs, since the excess of contribution margin over escapable costs helps to cover those (inescapable) costs that will be incurred regardless of the product-mix decision. The fixed costs must be separated into escapable and inescapable components as follows:

Product	Escapable fixed costs	Inescapable fixed costs	Total fixed costs
A	$550	$700	$1,250
B	410	450	860
C	380	120	500

We now see that shutting down product line B would lower the profits of the firm since only $410 of the $860 fixed expenses traceable to B would be avoided if product B were discontinued. Since product B generates a $780 contribution margin based on projected sales, it contributes $780 – $410 = $370 to cover inescapable fixed overhead and profit contribution. Product C, on the other hand, is generating only $360 in contribution margin, which is less than the $380 of fixed costs that could be avoided were it to be purged from the product line. Thus, the more detailed analysis of product C's fixed costs suggests that strong consideration be given to dropping this product. Based on escapable fixed costs, product C's sales would have to increase to 380/0.72 = 528 units (a 5.6% increase from current levels) before it could break even by covering its traceable and escapable fixed costs.

So far, we have analyzed the three products regardless of production constraints. Product A appears to be more profitable than product B since its unit contribution margin is $3.75, whereas product B's contribution margin is $2.60 per unit. Product A has higher escapable fixed costs than product B ($550 to $410), but its higher unit contribution margin enables it to break even at a volume of 147 units (550/3.75), whereas B's break-even quantity on escapable costs is 158 units (410/2.60). If we add the fact that sales of Products A and B could be expanded on a one-for-one basis (we can increase the sales of A by X units by decreasing the sales of B by X units and vice versa), we would apparently want to promote the sales of product A at the expense of Product B.

However, if we add a further complication that Products A and B share a common machine on which Product A requires 3 hours of processing per unit whereas Product B requires only 2 hours per unit, the situation becomes quite different. Under such conditions, Product B becomes more profitable since Product A generates $3.75/3 = $1.25 contribution margin per machine hour whereas Product B generates $2.60/2 = $1.30. Thus, Product B becomes the more profitable product to promote. In the limit, if we could completely offset any loss in Product A's sales with increased sales of Product B, we would attempt to run product A's sales down to zero, avoid $550 of escapable fixed costs, and produce and sell only B.

This example highlights the importance of carefully segregating both fixed from variable costs and escapable fixed costs from inescapable fixed costs, and determining when any production interdependencies exist among various products being considered. Basic knowledge of algebra enables us to incorporate the effects of the fixed versus variable and escapable versus inescapable cost classifications. However, when production interdependencies exist, we must turn, at least conceptually, to a mathematical programming formulation of the problem. Such a formulation permits us to find a product mix that maximizes contribution margin less escapable fixed costs with a feasible production schedule.

7.5. CVP ANALYSIS FOR A SINGLE PRODUCT IN MULTIPLE PERIODS

The capital-budgeting analysis is usually performed by comparing the initial cost of capital investment and the discounted (present worth of) future cash flow from the invested project. If the discounted cash flows exceed the initial investment, the project is worthwhile to be considered for investment.

Multi-period break-even analysis is a generalized approach to capital-investment decision-making. Its general model is derived as follows:

Let C = Initial cost (investment, or first cost) of a project considered for investment;
n = The useful life of the project;
S = Salvage value of the project after n years;
i = The cost of capital (interest rate);
P = Sales price of the product to be manufactured in this project (\$/unit);
V = Variable cost of the product (\$/unit);
F = Annual fixed cost of the project (\$/year);
Q = Annual production and sales volume (units/year);
Z = Net present worth or overall profits (before tax) of the project over its useful life (it may be written as $Z(Q)$ to indicate that it is a function of the annual production output).

If we assume that the cash flow occurs at the end of each year, the net present worth, $Z(Q)$, of the cash flows of the project over its n-year useful life can be expressed as follows:

$$Z(Q)=\left[(P-V)Q-F\right](P/A,\ i,n)+S(P/F,\ i,n)-C$$

where $(P - V)Q - F$ is the annual profit of the project ignoring its initial cost and salvage value.

To break even, as usual, $Z(Q)$ is set equal to zero and the equation is solved for Q, which is shown as Q_{BE}:

$$0=\left[(P-V)Q_{BE}-F\right](P/A,i,n)+S(P/F,i,n)-C$$

$$Q_{BE}=\frac{C-S(P/F,i,n)+F(P/A,i,n)}{(P-V)(P/A,i,n)}$$

or

$$Q_{BE}=\frac{C(A/P,i,n)-S(A/F,i,n)+F}{(P-V)}$$

The annual production output of Q_{BE} break evens the whole project where the cost of capital is i (in %) per year.

Of course, an annual output level (Q^*) necessary to obtain a targeted present value of annual profits (Z^*) can similarly be calculated:

$$Q_{BE}=\frac{\left[Z^*+C-S(P/F,i,n)\right](A/P,i,n)+F}{(P-V)}$$

or

$$Q_{BE} = \frac{\left(Z^* + C\right)\left(A/P,i,n\right) - S\left(A/F,i,n\right) + F}{\left(P - V\right)}$$

Example 7.13

Hilo Manufacturing Company is considering the purchasing of a new automatic machine for manufacturing its product. The machine has an initial cost of \$80,000, a useful life of 8 years, and a salvage value of \$10,000. If the machine is installed, only one skilled operator will be required to feed the machine at a cost of \$20 per hour. The hourly production of the machine will be 5 units. However, the annual maintenance and other fixed operating costs are expected to sum to \$5,000. The cost of capital is 12%. Now, the multi-period break-even analysis is used to answer the following two questions:

a) If the product is sold for \$10 per unit, how many units must the annual production be in order for the company to break even in this project?
b) If the production output is kept as 5,000 per year, at what price will the company break even in the project?

Solution

$$C = \$80{,}000,\ n = 8 \text{ years},\ S = \$10{,}000,\ i = 12\%,\ F = \$5{,}000,$$

$$V = (\$20/\text{hour})/(5 \text{ units/hour}) = \$4/\text{unit}$$

The overall profit (before taxes) of the company, in this project, is the net present worth of all cash flows, $Z(Q)$,

$$Z(Q) = \left[(P - V)Q - F\right](P/A, i\%, n) + S(P/F, i\%, n) - C$$

After substituting the corresponding values for the given variables, the CVP model will be:

$$Z(Q) = \left[(P - 4)Q - 5{,}000\right](4.9676) + 10{,}000(0.4039) - 80{,}000$$

or

$$Z(Q) = 4.9676(P - 4)Q - 100{,}799$$

Now, the questions are answered as follows:

(a) $P = \$10/\text{unit},\ Q_{BE} = ?$
To break even, set $Z(Q) = 0$, that is,

$4.9676(10 - 4)Q_{BE} - 100{,}799 = 0$
$Q_{BE} = \mathbf{3{,}382\ units}$

(b) $Q = 10{,}000$ units, $P_{BE} = ?$
To break even, again, set $Z(Q) = 0$:
$4.9676(P - 4)Q - 100{,}799 = 0$
$4.9676(P_{BE} - 4)(5{,}000) - 100{,}799 = 0$
$24{,}838(P_{BE}) - 200{,}151 = 0$
$P_{BE} = \mathbf{\$8.06}$

7.6. USEFULNESS AND APPLICATIONS OF CVP ANALYSIS

The cost-volume-profit (CVP) analysis is useful in finding answers to questions such as the following:

- What is the expected level of profit at a given sales volume?
- What sales amount is needed to cover all costs?
- What additional amount of sales is needed to achieve a desired level of profit?
- What additional sales amount must be made to meet the additional fixed costs to be incurred by a proposed plant expansion?
- What additional profit would be gained if the variable cost per unit is reduced by a given percentage?
- What will be the effect on profit due to a given percentage increase in sales?
- What increase in sales volume is needed to make up for a given reduction in sales price to maintain the current profit level?
- How much losses could the business incur if it faces a market downturn (sales loss)?
- How would the profits change if product prices are altered?
- What will be the production remaining capacity beyond the break-even point, which reflects the maximum amount of profit that can be achieved?
- What would the impact on the profit level if production automation (mainly a fixed cost) replaces labor (usually a variable cost)?

Decision-makers should routinely perform the CVP analysis to explore opportunities to lower the break-even point. A company can lower the break-even point by reducing fixed costs and increasing its contribution margin. Strategies to do so include:

- **Reducing Fixed Cost**: Manufacturing companies cut many fixed salaried positions and close plants that are not fully utilized. They outsource not-fully-utilized activities that involve fixed costs. Having lower fixed costs results in lower sales units required to cover them.
- **Reducing Variable Costs.** Lowering variable costs and expenses per unit increases contribution margin and, consequently, lowers the break-even point. Improper material processing or quality can result in unnecessary

waste and rework. Also, improper job assignments may translate into excessive labor costs.

- **Increasing Product Price**. Raising a product price is another method for increasing contribution margin and lowering the break-even point. Of course, price increase incurs some additional expenses associated with strategies such as increasing advertisement and promotions, enhancing the product quality, and improving the product packing and labeling. However, the decision-maker must be cautious that increased selling price may lower the sales units. Competition, contractual agreements, and government regulations may hinder the pricing strategy.

A decision-maker may be able to implement a combination of reduced fixed and variable costs, and moderately increased sales prices increases if possible. Some products might be redesigned to possess unique features that customers would be willing to pay for and the additional revenue is greater than the variable costs of the added features.

SUMMARY

Cost-Volume-Profit (CVP) analysis is a procedure to examine the effect of sales volume variations on costs and operating profit of a business. The analysis is based on the classification of expenses as variable (expenses that vary in direct proportion to sales volume) or fixed (expenses that remain unchanged over a relevant range, irrespective of the sales volume). Accordingly, the operating income (profit) is defined as follows:

$$\text{Operating Income}(Z) = \text{Sales }(PQ) - \text{Variable Costs }(VQ) - \text{Fixed Costs }(F)$$

where

Q = Sales volume (units of the product);
P = Sales price ($ per unit);
V = Variable cost per unit;
F = Fixed per period.

The CVP model can be expressed by the following formula;

$$Z = (P - V)Q - F$$

CVP analysis is used to determine the sales volume required to break even or achieve a specified profit level (also known as target profit). The cost-volume-profit analysis is an indispensable tool that helps in managerial, financial, and investment decisions. CVP analysis is used to answer questions such as:

- How changes in variable and fixed costs affect a firm's profit?
- How much must sell to break even (cover all costs) or reach a target profit?

- How will changes in selling prices affect profits?
- How will a change in the sales mix of products affect income?

CVP analysis makes several assumptions, including that the sales price, fixed, and variable costs per unit are constant.

The contribution margin is defined as sales minus variable costs. Therefore, the expression $(P - V)$ in the CVP formula is the contribution margin per unit of the product on which the CVP analysis is performed, and $(P - V)Q$ is the total contribution margin of selling Q units of the product generates. The contribution margin (CM) is usually stated as a contribution margin ratio (CMR) or percentage of sales, which is computed as follows:

$$\text{CMR}\left(\text{CM\%}\right) = \frac{\text{Contribution Margin}}{\text{Sales}} = \frac{(P-V)Q}{PQ} = \frac{P-V}{P}$$

CVP analysis can be used to determine a sales level required to achieve the desired profit (a target profit, Z^*). The sales level may be expressed in volume (number of units) or in dollar amounts. The required sales volume can be as follows:

$$\text{Sales Volume}\left(Q^*\right) = \frac{\text{Target Profit} + \text{Fixed Cost}}{\text{Contribution Margin Per Unit}} = \frac{Z^* + F}{P - V}$$

$$\text{Required Sales} = \frac{\text{Target Profit} + \text{Fixed Cost}}{\text{Contribution Margin Ratio}} = \frac{Z^* + F}{(P - V)/P}$$

The break-even point is reached when total costs and total revenues are equal, resulting in no gain or loss (i.e., operating income of $0). Businesses use the CVP analysis to determine the number of units of a product they need to sell at a given price point to break even, as follows:

$$\text{Break-even Units}\left(Q^*\right) = \frac{\text{Fixed Cost}}{\text{Contribution Margin Per Unit}} = \frac{F}{P - V}$$

$$\text{Break-even Sales Amount}\ (\$) = \frac{\text{Fixed Cost}}{\text{Contribution Margin Ratio}} = \frac{F}{(P - V)/P}$$

The break-even point computation is equivalent to finding the sales that yield a targeted profit of zero. Once the break-even point is reached, additional sales revenue is a profit. This CVP analysis is an essential tool in guiding managerial, financial, and investment decisions for current operations or future business plans.

Operating leverage is how much a firm's operating income reacts to a change in sales level. It is determined by a company's relative use of fixed versus variable costs. Businesses with high fixed costs relative to variable costs have high operating leverage. A business with high operating leverage will experience a sharp increase (decrease) in operating income with an increase (decrease) in sales level. The degree of operating leverage is determined by dividing the contribution margin by operating income.

Symbols and equations used in this chapter are listed in the following table:

Term	Symbol/Formula
Sales Units (Quantity)	Q
Sales (Selling) Price	P
Total Fixed Cost	F
Variable Cost Per Unit	V
Total Variable Cost	VC (or TVC) $= VQ$
Total Cost (Variable and Fixed Costs)	$\text{TC} = VQ + F$
Sales Amount (or Total Revenue)	$\text{TR} = PQ$
CVP (basic single-product model)	$Z = (P - V)Q - F$
Contribution Margin Per Unit	$\text{CMU} = (P - V)$
Contribution Margin (Total)	CM (or CMT) $= (P - V)Q$
Contribution Margin Ratio (or Percentage)	$\text{CMR} = \frac{\text{CM}_T}{\text{Sales}} = \frac{(P - V)Q}{PQ} = \frac{P - V}{P}$
Break-even Volume or Quantity (BEP in Units)	$Q_{BE} = \frac{F}{P - V}$
Break-even Value (BEP in Dollars)	BES = Total Cost $\text{BES} = P \cdot Q_{BE} = \frac{F}{\text{CMR}}$
Margin of Safety (in Dollars)	MOS = Sales – Break-even Sales
Margin of Safety Percentage	MOS % = 100(MOS/Sales Amount)
(Degree of) Operating Leverage	$\text{OL} = \frac{\text{CM}}{\text{Operating Income}}$

REVIEW QUESTIONS

R7.1. What are the four components of a single-product CVP model or analysis?

R7.2 How are **costs classified** in the income statement for CVP analysis?

R7.3. What is meant by a product's contribution margin ratio and how is it useful in planning business operations?

R7.4. How is the break-even point determined?

R7.5. What is the formula for computing break-even quantity?

R7.6. What is the formula for computing break-even sales amount (dollars)?

R7.7. How are the required sales to earn the target profit determined?

R7.8. What are the margin of safety and margin of safety ratio?

R7.9. Explain the term sales mix and its effect on break-even sales.

R7.10. What is the difference between the contribution margin ratio and the margin of safety ratio?

R7.11. Explain how to compute gross profit, operating profit, and net profit.

R7.12. How do changes in volume within the relevant range affect the break-even point?

R7.13. What important information is provided by the margin of safety as calculated in CVP analysis?

R7.14. The Creek-Water Company sells spring water. The company has an annual fixed cost of $450,000 but no variable cost. The projected demand (Q) – price (P) and marginal revenue (MR) functions for spring water are $Q=6{,}000-5P$ and $\text{MR}=1{,}200-0.4Q$. Determine the selling prices that maximize the company's annual profit.

PROBLEMS

P7.1. Assume that the CVP model for a company that makes a single product is

$$Z = \$48Q - \$48{,}000.$$

Required

a) How many units of the product must the company produce to break even?

b) How many units of the product must the company produce and sell in order to have a target profit of $90,000?

P7.2. KZ Company has provided the following data:

Sales Price Per Unit: $40
Variable Cost Per Unit: $15
Fixed Cost: $125,000
Expected Sales: 8,000 units

Required

a) What is the break-even point in sales dollars?

b) What is the expected margin of safety?

c) What is the operating leverage?

d) If the company wants to have a net income of $100,000, how many units must it sell?

P7.3. A company has provided the following data:

Sales Price Per Unit:	$80
Variable Cost Per Unit:	$48
Fixed Cost:	$240,000
Expected Sales:	10,000 units

Required

a) What is the break-even point in sales dollars?

b) What is the expected margin of safety?

c) If the company wants to have a net income of $144,000, how many units must it sell?

P7.4. Assume that the total cost, TC(Q), and demand functions for the product of a company are as follows:

$$TC(Q) = 200 + 60Q - 3Q2 + 0.2Q3 \quad [\text{Total Cost}]$$

$$Q = 100 - (1/3)P \quad [\text{Demand-Price relationship}]$$

where Q and P are the production (and sales) quantity and sales price, respectively.

Required

a) Find the maximum revenue.
b) Find the maximum profit.
c) Find the price that maximizes the profit.

P7.5. The K&K Company produces three products with price (P_i), variable cost (V_i), fixed cost (F_i), and machine time requirement for each product as shown below:

Product	P_i	V_i	F_i	Machine Hours per Unit
$i = 1$	$15	$6	$18,000	4
$i = 2$	25	10	36,000	6
$i = 3$	30	18	45,000	4

A maximum of 40,000 machine hours is available. Use contribution margin to answer the following questions:

Required

a) Which product is most profitable? Why?
b) If there were no machine time constraints, which product would have been most profitable? Why?
c) Assume that only 24,000 hours of supervisory time is available and supervision time required for products 1, 2, and 3 are 1 hour, 3 hours, and 2 hours, respectively. Formulate the problem for linear programming to maximize the profit.

P7.6. The DDT Company manufactures two products, A and B. The product-line income statement for the year 20X1 is given below:

DDT Company
Product-Line Income Statement
For the Year Ended December 31, 20X1

	Product A		Product B		
	Per Unit	Total Amount	Per Unit	Total Amount	Total ($1,000)
Sales	$15.00	$ 750,000	$22.50	$2,250,000	$3,000,000
Cost of Goods Sold:					
Materials	$4.00	200,000	$5.00	$500	$700,000
Direct Labor	2.00	100,000	3.00	300	400,000
Variable Overhead	1.50	75,000	2.50	250	325,000
Fixed Overhead	3.00	150,000	2.00	200	350,000
Cost of Goods Sold	$10.50	$525,000	$12.50	$1,250	$1,775,000
Gross Margin	$ 4.50	$ 225,000	$10.00	$1,000,000	$1,225,000
Fixed Sell and Adm. Exp.					325,000
Net Income Before Taxes					$900,000
Income Taxes (50%)					450,000
Net Income					$450,000

The sales mix is assumed to be the same at all volume levels from year to year. The management decided to increase the price of product A from $15.00 to $15.75 but to reduce the price of product B from $22.50 to $21.00 for the year 20X2 (the following year). At the same time, the company allocated $50,000 more for advertising for year 20X2, in which the material costs have dropped by 15% for both products but the direct labor increased by 10%. No other changes occurred in the year 20X2.

Required

a) Find the sales volume of each product to break even in 20X2.
b) What volume of sales for each product is required if the company is to earn a net income in 20X2 equals that earned in 20X1.

c) What volume of sales for each product is required if the company is to earn $300,000 net income after tax? Assume that the tax rate is 30%.

P7.7. The income statement of the Palmon Company is given below. Sales commissions are 5% of the sales amount. The company has a production capacity of 180,000 units per year. The results for 20XX were not satisfactory.

Palmon Company
Income Statement
For the Year Ended December 31, 20XX

Sales (100,000 units @ $5.00)			$500,000
Cost of Goods Sold:			
Direct Materials		$150,000	
Direct Labor		125,000	
Factory Overhead:			
Variable	$20,000		
Fixed	95,000	115,000	390,000
Gross Margin			$110,000
Operating Expenses:			
Selling Expenses:			
Variable:			
Sales Commission	$25,000		
Shipping	5,000	$30,000	
Fixed:			
Advertising, Salaries, etc.		45,000	$75,000
Administrative Expenses:			
Variable	$9,000		
Fixed	26,000	35,000	110,000
Net Income (or Loss)			$0

The Sales Manager has studied the market potential and believes that if the advertising expense is increased by $75,000 and the sales commission to 10% of the sales amount, the company will be able to sell the full production capacity without changing the sales price.

Required: Prepare the budgeted income statement, using a contribution margin format, and write the CVP formula for the company.

P7.8. A manufacturing company has two production machines that can be used 7 hours a day, 5 days a week for 50 weeks per year to manufacture its two products, A and B. The following is a summary of the relevant information about the products.

	Product A	Product B
Selling price (per unit)	\$60	\$35
Variable cost (per unit)	\$30	\$10
Units that can be manufactured per machine hour on the same production line	10	6
Estimated demand (units)	15,000	22,000

Required: To maximize its profit, how many machine hours should the company devote to the manufacturing of each product?

P7.9. The Maldust Company manufactures and sells a product that has a variable cost of \$240. The company's fixed costs are \$169,000 per month. Based on the historical data and marketing research, the company has developed the following price–demand relationship:

$$Q = 375 - 0.05P,$$

where P is the selling price and Q is the demand quantity.

Required

a) Develop a CVP model for the company.
b) Determine the break-even quantity(ies).
c) Determine the selling price that maximizes the company's monthly profit.
d) Compute the company's monthly maximum profit.

P7.10. Bullarson Corporation produces and sells a single product. Information on its costs follow:

Variable Costs:	
Production	\$4 per unit
Sales, General and Administration	\$3 per unit
Fixed Costs:	
Production	\$18,000 per month
Sales, General and Administration	\$15,000 per month

Required

a) Assume the company produced and sold 15,000 units. At this level of production and sales, it produced a profit of \$45,000. What was the company's sales price per unit?

b) The projected production and sales in the upcoming month are 16,000 units. The variable costs per unit and the total fixed costs are expected to be the same as in the current month. However, the company sets its sales price at $12.50 per unit. What is the projected margin of safety and the degree of operating leverage for the next month?

P7.11. Sambo Manufacturing has annual fixed costs of $500,000 and is producing and selling a single product. Its projected sales are 150,000 units. The management desires an after-tax net income of $90,000. The company's income-tax rate is 25%. What is the maximum amount that the company can incur as variable costs per unit and still meet its target profit if the sales price is estimated at $10?

P7.12. The Maldar Company manufactures and sells a single product and projects the following financial information for the upcoming year:

Sales	60,000 units
Variable Costs	$6.00 per unit
Fixed Costs	$335,000 per year
Unit Sales Price	$15.00

Required

a) How many units would the company have to sell to earn a profit before taxes of $250,000?

b) If Maldar Company achieves its desired, what will be its degree of operating leverage?

Index